Emiliano **Lepore**
Nicola **Pugno**

An Experimental Study

on Adhesive or Anti-adhesive, Bio-inspired Experimental Nanomaterials

VERSITA

Versita Discipline:
Engineering, Industry, Transportation

Managing Editor:
Elisa Capello

Language Editor:
Mary Boyd

Published by Versita, Versita Ltd, 78 York Street, London W1H 1DP, Great Britain.

ISBN (paperback): 978-83-7656-080-9

ISBN (hardcover): 978-83-7656-081-6

ISBN (for electronic copy): 978-83-7656-082-3

Managing Editor: Elisa Capello

Language Editor: Mary Boyd

www.versita.com

Cover illustration: © Emiliano Lepore

"The only way to do great work is to love what you do. If you haven't found it yet, keep looking. Don't settle. As with all matters of the heart, you'll know when you find it. And, like any great relationship, it just gets better and better as the years roll on. So keep looking until you find it. Don't settle. [...] Stay hungry. Stay foolish"
(S. Jobs, an American computer entrepreneur and innovator)

"Genius is one percent inspiration, ninety-nine percent perspiration"
(T. Edison, an American inventor and entrepreneur)

"Poor is the pupil who does not surpass his master"
(L. da Vinci, an Italian Renaissance polymath)

"I seek not the answer but to understand the question"
(H. Kroto, a British Nobel Prize Winner)

Dedicated to
small things which become great

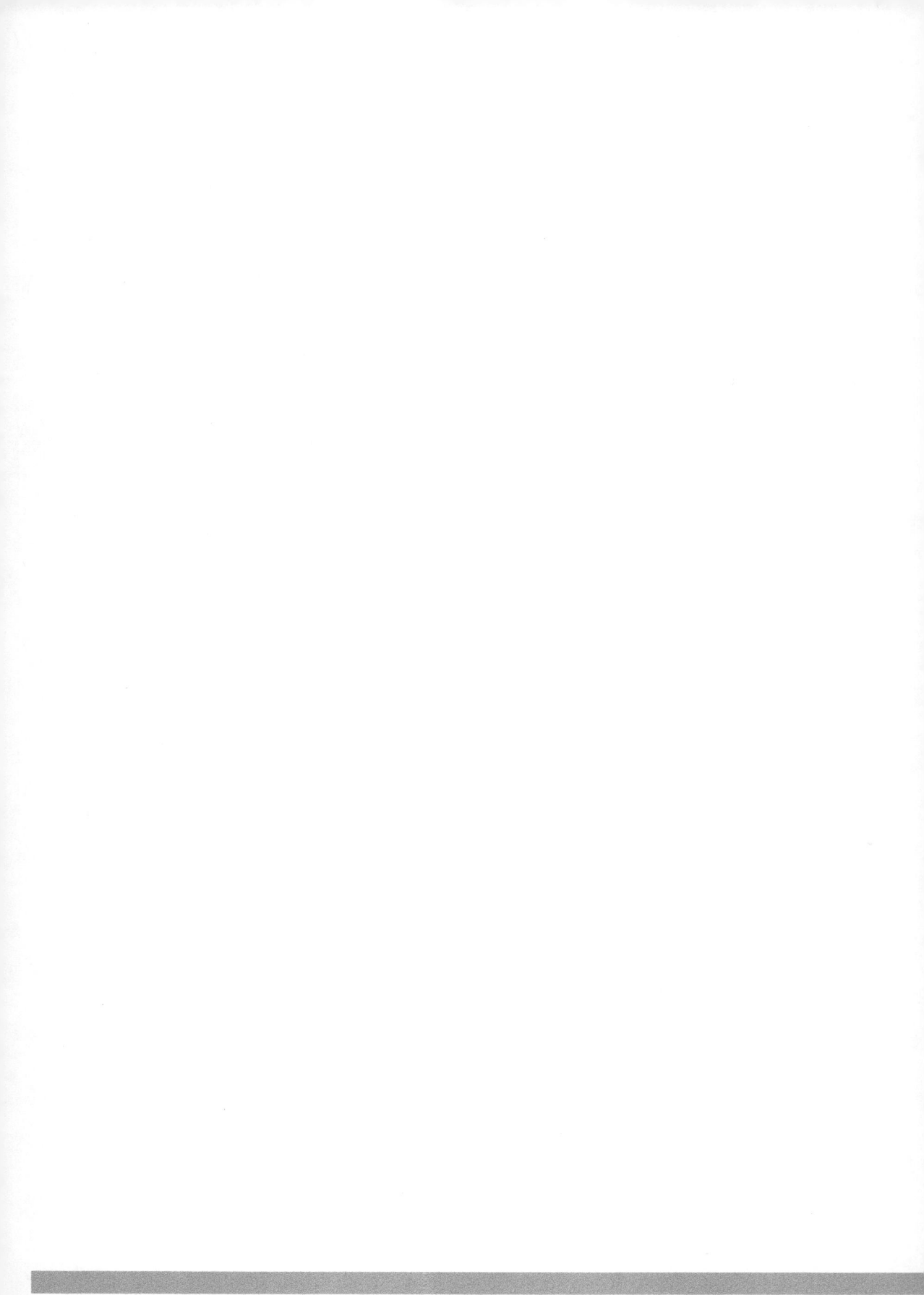

Contents

Acknowledgements

First and foremost, I would like to express my deep gratitude to my tutor Prof. Nicola Pugno for his theoretical rigour and fascinating scientific approach. I thank Nature for its continuous stimulus and inspiration to optimization.

I wish to sincerely thank my wife Maria, who has always encouraged me and believed in my abilities, showing me every day how small wonderful things could become great, only with passion.

I am very thankful to my mother Marilena and my father Vittorino who, when I was a child, taught me how only honesty and respect for the others are the two milestones in human relationships, thus also in work. Unfortunately, these are not common human values.

List of Figures

Figure 1.10 Interpretation of experimental results for adhesion tests on various roughness of PMMA surfaces. (A) Setae (represented in blue) and sub-hierarchical structures (spatulae) can adapt well on virgin PMMA; (B) The adhesion on PMMA2400 is better because of the higher number of spatulae-substrate interactions; (C) On PMMA800, only partial contact interactions are achieved.

Figure 2.1 Spider and gecko feet showed by SEM (Zeiss EVO 50). The Tokay gecko (Fig.2.1F) attachment system is characterized by hierarchical hairy structures, which start with macroscopic lamellae (soft ridges ~1 mm in length, Fig.2.1H) and branches into setae (30-130 µm in length and 5-10 µm in diameter, Fig.2.1I, 2.1L (Hiller, 1968; Ruibal, 1965; Russell, 1975; Williams, 1982)). Each seta consists of 100-1000 substructures called spatulae, the contact tips (0.1-0.2 µm wide and 15-20 nm thick, Fig.2.1M (Hiller, 1968; Ruibal, 1965)) responsible for the gecko's adhesion. Terminal claws are located at the top of each singular toe (Fig.2.1G). Van der Waals and capillary forces are responsible for the resultant adhesive forces (Autumn, 2002a; Sun, 2005a), whereas the claws guarantee an efficient attachment system on surfaces with very large roughness. Similarly, an analogous ultrastructure is found in spiders (*e.g. Evarcha arcuata* (Kesel, 2003)). Thus, in addition to the tarsal claws, which are present on the tarsus of all spiders (Fig.2.1C), adhesive hairs can be distinguished in many species (Fig.2.1D, 2.1E). Like for insects, these adhesive hairs are specialised structures that are not restricted only to one particular area of the leg, but may be found either distributed over the entire tarsus - as for lycosid spiders - or concentrated on the pretarsus as a tuft (scopula) situated ventral to the claws (Fig.2.1A, 2.1B) - as in the jumping spider *Evarcha arcuata* (Kesel, 2003).

Figure 2.2 Weibull statistics (F is the cumulative probability of detachment/failure and *ti* is the measured adhesion time) applied to the measured adhesion times on PMMA surfaces. PMMA 1 (red lines, for which we made 4 sets of measurements on four different days with gecko G1), PMMA 2 (dotted lines, for which we made 2 sets of measurements on two different days, one with gecko G1 (red) and one with gecko G2 (blue)), and PMMA 3 (blue double-line, for which we made the measurements in a single day with gecko G2).

Figure 2.3 A simple interpretation of our experimental results on the adhesion tests for living geckos on PMMA surfaces with different roughness. (A) Setae cannot adapt well on PMMA 1; (B) On PMMA 2, the adhesion is enhanced because of the higher compatibility in size between setae and roughness; (C) On PMMA 3 only partial contact is achieved. On the right, we report the analysed three-dimensional roughness profiles for all three investigated surfaces (from the top: PMMA 1, 2 and 3).

Figure 3.1 The Tokay gecko's adhesive system was observed by FESEM (Zeiss SUPRA 40) (A, B) and by SEM (Zeiss EVO 50) (C, D). (A) Toe and FESEM micrograph of the setae (B). SEM micrograph of the setae (C) A nanoscale array of hundreds of spatulae (D).

Figure 3.2 The Tokay gecko's adhesive system was observed by FESEM (Zeiss SUPRA 40). (A) The Tokay gecko's toe. (B, C) The connection area between adjacent lamellae, localized perpendicular to the longitudinal axis of each digit, is covered by nanostructured hairy units; (D) at high magnification.

Figure 3.3 The Tokay gecko's adhesive system was observed by FESEM (Zeiss SUPRA 40). (A) The Tokay gecko's toe. (B, C) The edge of the gecko's toe is covered by nanostructured hairy units; (D) at high magnification.

Figure 3.4 The experimental Tokay gecko with adherent elastic cloth bandaging and metallic connection hook on the measurement platform.

Figure 3.5 Force-displacement measurement platform.

Figure 3.6 Normal adhesive force-displacement curves on PMMA surfaces after the first and second moults. Snapshots show five specific instants of the gecko's displacement at 0, 148, 273, 423, and 723 g of hanging weight (W is the applied weight, W_G is the gecko's weight, δ is the gecko's displacement, and δ_{MAX} is the gecko's maximum displacement).

Figure 3.7 Normal adhesive force-displacement curves on glass after the first and second moults. Snapshots show five specific instants of the gecko's displacement at 0, 148, 348, 423, and 648 g of hung weight (W is the applied weight, W_G is the gecko's weight, δ is the gecko's displacement, and δ_{MAX} is the gecko's maximum displacement).

Figure 3.8 AFM characterization of the PMMA surface.

Figure 3.9 AFM characterization of the glass surface.

Figure 3.10 Damage imposed by the adhesive tests: (A) Diffused inflammation of gecko toes; (B) The gecko's healthy foot, for comparison; (C) Small, thin wound located on the gecko's skin between one toe and the next.

Figure 4.1 (A) A schematic 3D representation of the measured angle between the opposing front and rear feet (β_F) and between the first and fifth toe (β_T) of each foot on inverted surfaces (inset adapted from Y. Tian, N. Pesika, H. Zeng, K.

Rosenberg, B. Zhao, P. M., K. Autumn, and J. Israelachvili, *Adhesion and friction in gecko toe attachment and detachment*, 19320-19325, PNAS, December 19, 2006, vol. 103, no. 51; Copyright (2006) National Academy of Sciences, U.S.A.). The Tokay gecko adhesive system observed using FESEM (Zeiss SUPRA 40) (B, C, D) and by SEM (Zeiss EVO 50) (E). The gecko's toe (B), FESEM micrograph of setae arrays (C), SEM micrograph of several setae (D) and nanoscale array of hundreds of spatula tips (E).

Figure 4.2 The measured angle f_F between the opposing front and rear feet on different surfaces (steel, aluminium, copper, PMMA, and glass).

Figure 4.3 The measured angle β_T between the first and fifth toe: on the aluminium surface for all legs (A), or for the FR leg on different surfaces (steel, aluminium, copper, PMMA, and glass).

Figure 4.4 From the multiple peeling theory (Sitti, 2003), the dimensionless force f *versus* adhesion angle α using experimental mean values for α_F and α_T (fitting parameters λ reported in Table 4.1).

Figure 5.1 Side (A) and top (B) view of the centrifugal machine used to measure the insects sSF (M1: passive rotating linchpin; M2: electric motor connected to M1 with a transmission belt; FC: frequency controller to set the M2 rotational speed; RA: rotational axis; C: camera; B1: external box; B2: internal small box where specimens were placed in; M: middle of the internal box; L: lamp; BC: bicycle computer; CW: counterweight).

Figure 5.2 Two subsequent frames from a video: before detachment, the insect stands still on the surface (A) and, one frame later, it is in the box corner (B). These frames are extracted from a preliminary video without the use of the small box (B_2).

Figure 5.3 AFM characterization of the (A) steel, (B) aluminium, (C) copper, and (D) Cp surfaces.

Figure 5.4 The adhesive structures of the legs of *Blatta Orientalis*. (a) Frontal and (b) lateral view of a leg and some detailed micrographs (c, d, e, f, g) (*d* is the claw tip diameter).

Figure 5.5 The sSF for each individual are grouped by surfaces.

Figure 6.1 FESEM microscopies of the tested PS surfaces.

increased in 2-µl increments, from 2-µl up to the minimum sliding volume (SV) of the droplet, at which sliding occurs (final step, *n*). The sliding speed (SS) was determined by measuring the time required to cover a fixed distance of 10 mm (mean velocity).

Figure 7.3 Details of: (a) fresh lotus leaf (LL), (f) lotus leaf resulted after copying process (CLL), (m) negative copy (C1), and (r) positive copy (C2). In particular, b, g, and s (n) show randomly distributed convex (concave) cell papillae; c, h, and t (o) show magnified detail of the convex (concave) cell papillae; wax tubules are magnified in d (natural wax tubules) and i (the wax tubules are broken due to C1 deposition and peeling). The nano-tubules are absent on C1 and C2. Water droplet on the surface of: (e) fresh lotus leaf (LL), (l) lotus leaf resulted after copying process (CLL), (p) negative copy (C1), and (u) positive copy (C2). (q) and (v) show the shape of a water droplet on C1_control and C2_control surfaces, respectively. For LL and CLL, no control surface can be defined. The measurements reported in e, l, p, u, q, v are the average CA ± st.dev..

Figure 8.1 FESEM image of the spinnerets of *Meta menardi* (1. Anterior lateral; 2. Posterior median; 3. Posterior lateral).

Figure 8.2 Egg sac of the spider *Meta menardi*. Photo by Francesco Tomasinelli (2009).

Figure 8.3 Distinction of the stalk types: cable-like (Group A) (a) and ropey (Group B) (b).

Figure 8.4 FESEM characterization of the silk stalk at different magnifications.

Figure 8.5 Stress-strain curves for group A (a) or B (b) stalks.

Figure 8.6 Weibull statistics for stress of group A (a) and B (b) stalks.

Figure 8.7 Detailed views of fracture surfaces of broken silk fibers.

Figure 8.8 FESEM characterization of the stalk cut with FIB: (a, b, c) at an eye angle of 52° and (d) from the top.

Figure 8.9 The maximum toughness of different types of (mainly spider) silks.

Figure 8.10 The maximum strength of different types of (mainly spider) silks.

Figure 8.11 The maximum strain of different types of (mainly spider) silks, showing the record for egg sac silk stalks observed in the present study.

List of Tables

General Introduction

This experimental monograph presents the results of research performed in five different facilities: the Laboratory of Bio-inspired Nanomechanics "Giuseppe Maria Pugno" at the Politecnico of Torino; the "Nanofacility Piemonte" at the INRIM Institute in Torino; the Division of Dental Sciences and Biomaterials of the Department of Biomedicine at the University of Trieste; the Physics Department of the Politecnico of Torino; the Toscano-Buono Veterinary Surgery in Torino and the Department of Human and Animal Biology at the University of Torino.

The adhesive abilities of insects, spiders, and reptiles have inspired researchers for a long time. All these organisms present outstanding performance particularly for force, surface adhesion, and climbing abilities for their size and weight. Scientists have focused on the gecko's adhesive paw system and climbing abilities, and its adhesion mechanism has been an important topic of research for nearly 150 years. However, certain phenomena exhibited by geckos are still not completely understood and represent the main challenge of scientific discussion aiming to better understand the gecko's adhesive ability. The Tokay gecko (*Gekko gecko*) is the most studied gecko among more than 1050 Gekkonid lizard species in the world due to its strong adhesive ability. Because this study reports clear experimental measurements on two living Tokay geckos, it is comparable to scientific results reported in the literature.

This monograph first discusses the influence of surface roughness on the gecko's adhesion on inverted surfaces of Poly(methyl meth-acrylate) (PMMA) and glass in **Chapter 1** and of PMMA with different surface roughness in **Chapter 2**. **Chapter 3** describes the gecko's maximum normal adhesive force, and **Chapter4** explores the optimal adhesion angle at different hierarchical levels.

In addition, it is well known how small insects can carry many times their own weight and walk quickly, but another interesting ability is their extremely high surface adhesion. In recent decades, many scientists have studied a number of insects in order to better understand and measure their adhesive abilities. Biological adhesion can be obtained through different adhesive mechanisms (*e.g.* claw, clamp, sucker, glue, friction). In particular, **Chapter 5** of this monograph focuses on living specimens of the non-climbing cockroach (*Blatta Orientalis* Linnaeus) and evaluates their maximum shear safety factor on artificial surfaces using a centrifuge machine.

In general, an animal's adhesive structure and mechanism can be correlated to the micro-structured roughness of natural substrata (*e.g.* plant surfaces), which animals encounter in their natural environment.

In nature, plants exhibit an extraordinary variety of morphologies and surface structures. Some plants possess two special properties: superhydrophobicity (or water-repellency) and self-cleaning (or dirt-freedom). These two related phenomena were observed for the first time by Aristotle more than 2,000 years ago, but it was only in the 20th century that scientists examined them on natural leaves, *e.g.* the lotus (*Nelumbo nucifera*) on which "raindrops take a clear, spherical shape without spreading, which probably has to be ascribed to some kind of evaporated essence", as Goethe described in 1817. Scientific literature indicates a strong influence of surface roughness on wettability and self-cleaning behaviour. This well-defined problem is of particular interest to the Indesit Company since it is desirable to find an industrial solution to leave the internal sides of refrigerators, which are composed of polystyrene, clean from condensed water or dirt. A collaborative project was initiated to understand the role of surface roughness on these two properties and produce a superhydrophobic and self-cleaning surface. Two industrial processes, plasma and thermoforming treatments, were applied to polystyrene surfaces. The influence of these industrial treatments on surface wettability are analysed in **Chapter 6**. **Chapter 7** of this monograph reports the method with which an artificial biomimetic superhydrophobic polystyrene surface, copying a natural lotus leaf, was designed.

Additionally, spider silks display superior mechanical properties but only in the last few decades have researchers studied various types of silks and evaluated their different mechanical properties. It is well-known that the mechanical behaviour of spider silks varies according to type, and silk properties have been demonstrated to be species-specific and linked to silk-based peptide fibrils and protein aggregates with varying structural and mechanical properties. The dragline silk (or radial silk) and flag silk (or circumferential silk) of orb weaving spiders have been characterized in scientific literature, whereas few studies have been conducted on bundles, which connect the cocoons of *Meta menardi* to the ceilings of caves. **Chapter 8** describes the testing of these materials to assess their mechanical properties, including stress, strain, and toughness. The nanometer scale characterises this monograph and its subject matter, including gecko spatulae, the waxy nanotubules of the lotus leaf, and the fibroin protein materials which constitute spider silks.

Adhesive Materials

Chapter 1

The Weibull Statistics Applied to the Adhesion Times of Living Tokay Geckos on Nanorough Surfaces

Abstract

In this chapter we demonstrate that living tokay geckos (*Gekko gecko*) display adhesion times following Weibull Statistics. We have considered two different geckos, male or female, adhering on different surfaces, glass or Poly(methyl meth-acrylate) (PMMA) with different roughness. We have performed detailed surface topography characterizations by means of a three-dimensional optical profilometer. The analysis suggests the existence of a "weakest link" in the gecko's adhesion and is able to quantify its degree of brittleness for different surface roughness.

1.1. Introduction

In the world, there are more than 1050 species of geckos divided into 50 families. The Tokay gecko (*Gekko gecko*) is the second largest gecko species: an individual can weigh up to 150-200 grams. The gecko's climbing ability has attracted human attention for more than two millennia. The gecko's ability to "run up and down a tree in any way, even with the head downwards" has been observed since the time of Aristotle (Aristotle, 343), who mentioned these curious creatures in his manuscript, *Historia Animalium*, written four centuries before Christ.

Until the mid-twentieth century, scientific observations had not permitted a thorough understanding of the capacity of the gecko to stay stuck motionless or running on vertical or inverted surfaces (Simmermacher, 1884; Schmidt, 1904; Dellit, 1934; Ruibal, 1965). Only after the electron microscopy's development in the 1950s were researchers able to note the hierarchical, from the nano- to the macro-scale, morphology of the gecko's feet (Autumn, 2006a; Russell, 1975; Russell, 1986; Schleich, 1986; Gennaro, 1969). A Tokay gecko's foot consists of hierarchical structures (Fig.1.1) starting with macroscopic lamellae (soft ridges ~1 mm in length), from which branch off setae (30-130 μm in length and 5-10 μm in diameter). Each setae terminates with 100-1000 substructures such as spatulae (0.1-0.2 μm wide and 15-20 nm thick), responsible for the gecko's adhesion. More recently, numerous studies (see (Hiller, 1968; Autumn, 2002a; Autumn, 2007; Autumn, 2008; Arzt, 2003; Autumn, 2002b; Bergmann, 2005; Huber, 2005a; Autumn, 2006b; Autumn, 2006c; Autumn, 2000; Huber, 2005b) and related references) explain the factors that allow the gecko to adhere to and detach from surfaces. Very recently, van der Waals attraction (Autumn, 2000) and capillarity (Huber, 2005b) have been recognized as the key mechanisms in the gecko's adhesion.

Like geckos, many other creatures such as beetles, flies and spiders possess the remarkable ability to move on vertical surfaces and ceilings (e.g. see (Stork, 1980; Dai, 2002) and related references). Their adhesive abilities arise from the micro/nanostructures that compose their attachment pads. It is noteworthy that, as the mass of the creature increases, the size of the terminal attachment elements decreases and their densities increase (Arzt, 2003) to enhance the adhesion strength. Thus, moreso than insects and spiders, geckos exhibit the most versatile and effective dry adhesion known in nature, as imposed by their larger mass. Mimicry of the gecko's adhesion could lead to a revolution in material science (Yurdumakan, 2005; Haeshin, 2007; Lepore, 2008; Pugno, 2008a; Pugno, 2008b; Lepore 2010) or even the invention of Spiderman suits (Pugno, 2007a).

In this study, we report new observations on the adhesion times of living tokay geckos following two different *in vivo* experiments. We have considered two different geckos, one male and female, adhering on different surfaces, glass and Poly(methyl meth-acrylate) (PMMA) with different roughness. All surfaces were previously analysed with a three-dimensional (3D) optical profilometer. The data have been treated using Weibull statistics, showing a relevant statistical correlation.

Although the measurement of failure time is an interesting parameter, it cannot be directly correlated with the force and energy values of prior studies. Moreover, since our data are from live geckos, the role of the animal's behaviour in failure time cannot be *a priori* excluded and the adhesion times have to be considered as indicative of the entire biosystem, *i.e.*, not only of the animal's adhesive ability but, for example, also of muscular fatigue (it is well-known that geckos must produce shear forces to maintain adhesive forces (Autumn, 2006b). Given the long attachment times, it is possible that the geckos might have become fatigued, limiting their clinging ability). Nevertheless, the extraordinary adhesive ability that we have observed after moulting suggests that the measured adhesion times are mainly linked to adhesive ability and are scarcely influenced by other factors, such as muscular fatigue.

1.2. Surface Characterization

The characterization of PMMA and glass surfaces was performed with a three-dimensional (3D) optical profilometer, Talysurf CLI 1000, equipped with the CLA Confocal Gauge 300 HE (300 µm range and 10 nm vertical resolution), both from Taylor Hobson, Leicester, UK. The parameters tuned during the analysis were the measurement speed (50 µm/s), the sampling rate (100 Hz), the measured area (0.1 x 0.1 mm²), the resolution in "*xy*" plane (0.5 µm), leading to a final resolution of 201 points/profile. All parameters were referred to a 25 µm cut-off.

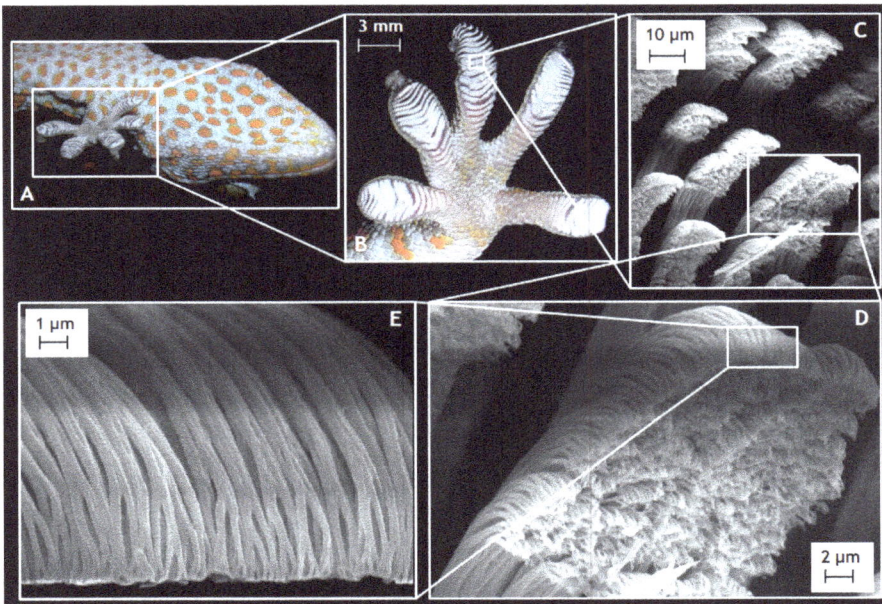

Figure 1.1 The gecko's hierarchical adhesive apparatus. (A) Ventral view of the Tokay gecko (*Gekko gecko*). (B) Gecko's foot. Scanning electron microscope (SEM) micrographs of (C) the setae, (D) at higher magnification, (E) terminating in hundreds of spatula.

The roughness parameters of interest were: the standard amplitude parameters *Ra*, *Rq*, *Rp*, *Rv*, *Ssk* and the hybrid parameters *Sdr* (for details, see Fig.1.2). *Ra* represents the arithmetical average roughness ($R_a = \dfrac{1}{l_n} \int_0^{l_n} |y(x)| dx$); *Rq* is the mean square roughness and represents the mean square deviation of the profile from the middle line ($R_q = \sqrt{\dfrac{1}{l_n} \int_0^{l_n} y^2(x) dx}$); *Rp* and *Rv* are, respectively, the height of the highest peak and the depth of the deepest valley (absolute values). Sa, Sq, Sp, Sv have the same meaning of Ra, Rq, Rp, Rv, but refers to surfaces and not to areas. The parameters Ssk and Sdr offer a comprehensive overview of the surface's characteristics, indicating, respectively, the surface skewness and the surface complexity. When Ssk is close to 0, the surface is equally distributed on the middle plane (p_m), when lower than 0 the surface is characterized by plateaus and several deep thin valleys, whereas when higher than 0 the surface is characterized by plateaus and several peaks. The parameter Sdr compares the effective surface (l_e) with the nominal one (l_n): when close to 0%, the surface is smooth, when higher the surface is characterized by a specific superficial complexity.

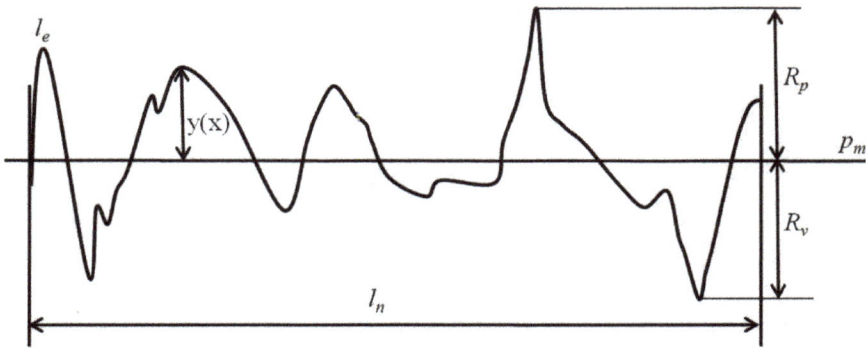

Figure 1.2 General scheme of a profile for the definition of the roughness parameters.

Virgin PMMA and glass surfaces, tested in the first experiment, present nearly homogeneous roughness without significant anomalous alterations, apart from small isolate bubbles on the surface of glass derived from melting during the fabrication process. Figs.1.3 and 1.4 show the virgin PMMA and glass surface (A) three-dimensional topographies and (B) two-dimensional profiles. PMMA surfaces with different roughness, namely PMMA2400 or PMMA800, have also been considered. PMMA2400/800 surfaces are obtained by a manual process that consists of clockwise circular movement for 2 minutes on the material sample using sandpaper 2400/800. Figs.1.5 and 1.6 show the PMMA2400/800's surface (A) topographies and (B) profiles. We note that the roughness parameters allow us to appreciate the differences between virgin PMMA and glass surfaces and, more importantly, become nearly one order of magnitude greater for machined PMMA surfaces (with the exception of the skewness that changes sign). Table 1.1 summarizes average roughness parameters of the characterized surfaces.

Table 1.1 Roughness parameters of the characterized surfaces.

	Glass	PMMA	PMMA2400	PMMA800
S_a (µm)	0.031 ± 0.0019	0.033 ± 0.0034	0.481 ± 0.0216	0.731 ± 0.0365
S_q (µm)	0.041 ± 0.0034	0.042 ± 0.0038	0.618 ± 0.0180	0.934 ± 0.0382
S_p (µm)	0.366 ± 0.1649	0.252 ± 0.0562	2.993 ± 0.1845	4.62 ± 0.8550
S_v (µm)	0.434 ± 0.2191	0.277 ± 0.1055	2.837 ± 0.5105	3.753 ± 0.5445
S_{sk}	-0.381 ± 0.4630	-0.122 ± 0.1103	0.171 ± 0.1217	0.192 ± 0.1511
S_z (µm)	0.609 ± 0.2791	0.432 ± 0.1082	4.847 ± 0.2223	6.977 ± 0.2294
S_{dr} (%)	0.574 ± 0.0724	0.490 ± 0.0214	15.1 ± 1.6093	28.367 ± 2.2546

A

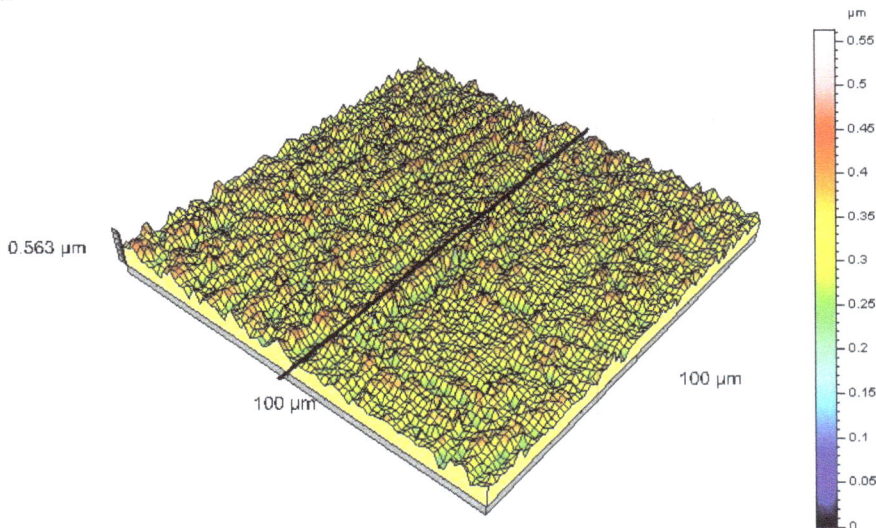

B

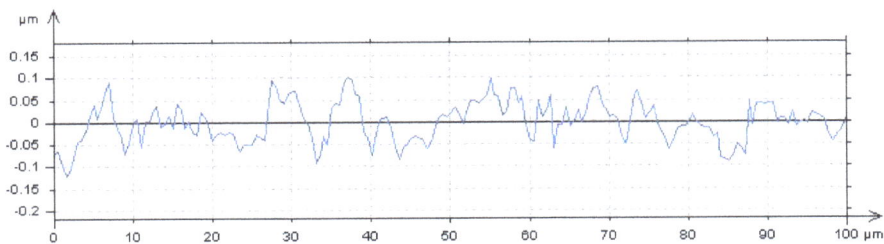

Figure 1.3 PMMA virgin surface. (A) Three-dimensional topography. (B) Two-dimensional profile (extracted at 50 µm from the edge of the square measured area).

A

B

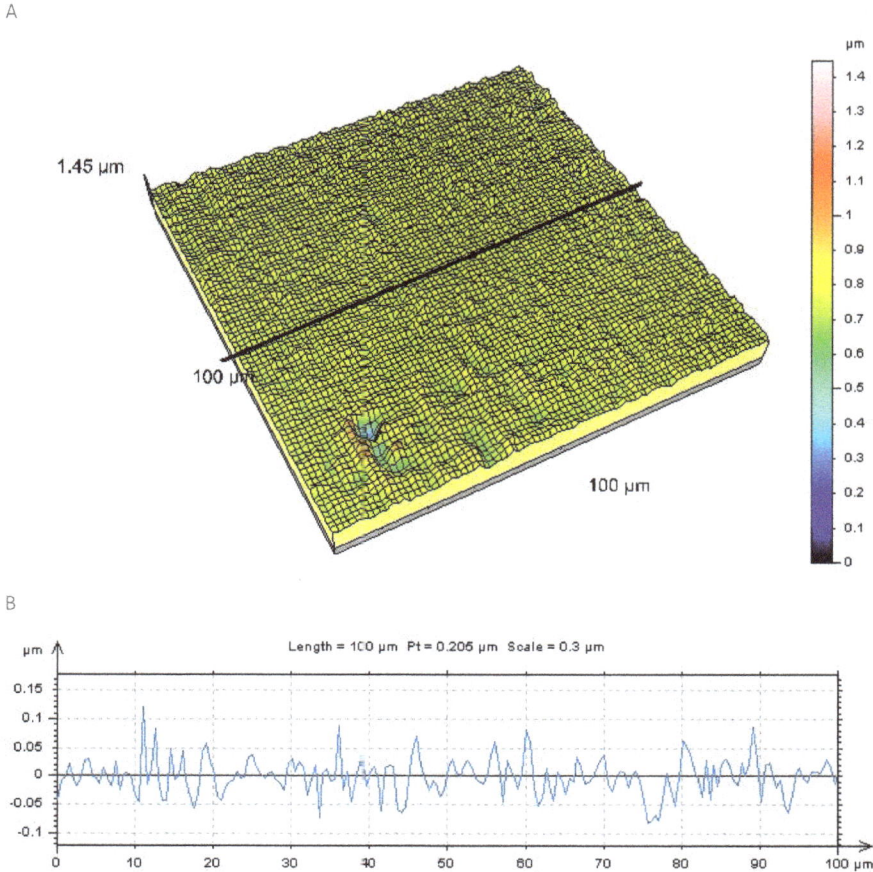

Figure 1.4 Glass surface. (A) Three-dimensional topography. (B) Two-dimensional profile (extracted
at 50 μm from the edge of the square measured area).

For the first experiment, the roughness parameters *Sa* and *Sq* did not
allow appreciation of significant differences between virgin PMMA and glass
surfaces. In particular, the parameter *Sa* represents the area between the
middle line and the roughness profile, but it is a measurement which cannot
distinguish the difference between solid areas and voids. For this reason it
could not provide information about the superficial design and the skewness
of the surface. However, in the second experiment the roughness parameters
Sa and *Sq* of PMMA2400 and PMMA800 revealed the first important difference
between machined PMMA surfaces and virgin PMMA: these two surface
parameters present a value one order of magnitude higher than those of virgin
PMMA.

The present study focuses on the parameters Sp, Sv and Sz, combined with the parameter Ssk and the roughness parameter, Sdr because of their significance in surface characterization. The roughness parameters, Sp and Sv, denote that virgin PMMA and glass have similar values for the highest peak and deepest valley (magnitude of hundreds of nm). The same observation can be made for PMMA2400 and PMMA800, but for the latter surface the values of Sp and Sv are of a few µm (~3 and ~4 µm, respectively), one order of magnitude higher than those of virgin PMMA and glass.

When we consider the Sp and Sv parameters, the virgin PMMA surface shows lower values than glass and, as expected, PMMA2400 presents lower values compared to PMMA800. However, both virgin PMMA and glass have higher values for the Sv parameter compared to the respective value of Sp. This indicates a surface characterized by higher depth (absolute value) of valleys compared with the height of peaks. As a consequence of the manual process of surface manufacture, both PMMA2400 and PMMA800 denote a reverse trend, showing a surface with higher height of peaks in comparison with the depth (absolute value) of valleys.

Moreover, an analysis of the Sz parameter reveals the spatial distribution of the five highest peaks and five deepest valleys. The virgin PMMA surface is characterized by less marked peaks and valleys compared to the glass surface ($Sp_{-SmPMMA} < Sp_{-Glass}$, $Sv_{-SmPMMA} < Sv_{-Glass}$), but on the virgin PMMA surface the five highest peaks are nearer to one another ($Sz_{-SmPMMA} = 0.432$ µm) than those of the glass surface ($Sz_{-Glass} = 0.609$ µm). Both the virgin PMMA surface and glass surface show negative Ssk, indicating a trend of plateau and several deep thin valleys, which support our findings about the Sp and Sv parameters. This trend is more marked on the glass surface than on the virgin PMMA surface. PMMA2400 and PMMA800 possess positive Ssk values, which indicate a trend of plateau and several peaks in agreement with the previously reported results for the Sp and Sv parameters.

In conjunction with Ssk, Sdr describes the complexity of the surface. Both virgin PMMA and glass are characterized by small values of the Sdr parameter ($Sdr < 1$ %), indicating that both surfaces have a reduced superficial complexity; the maximum value, obtained for the glass surface, indicates that the measured area of the curvilinear surface exceeds the area of the support surface by a maximum factor of 0.574%.

In contrast, the two machined surfaces (PMMA2400 and PMMA800) present a high superficial complexity, so the real measured area exceeds the support scanning area by a factor of ~15 % and ~28 %, respectively.

Figs.1.3B and 1.4B present the characteristic profiles of virgin PMMA and glass analysed in the first set of experiments. These profiles confirm our previously reported findings in that both surfaces have mainly plateau and several deep thin valleys. This feature seems to be more marked on the glass profile than on the virgin PMMA profile. During our analysis, extracted profiles enabled us to make some further observations. The glass profile is characterized by a higher irregularity compared to the PMMA profile; it shows very deep thin valleys close to very marked peaks. On the contrary, in the distribution of valleys and peaks, virgin PMMA presents a profile that is nearer to the middle line; the height of peaks and valleys is of the same order of magnitude compared to the glass profile's height, while the lateral width is almost one order of magnitude higher than that of the glass profile.

Figs.1.5B and 1.6B present the characteristic profiles of machined surfaces (PMMA2400 and PMMA800) analysed in the second set of experiments and compared to virgin PMMA. Their superficial profiles are similar and show a regular

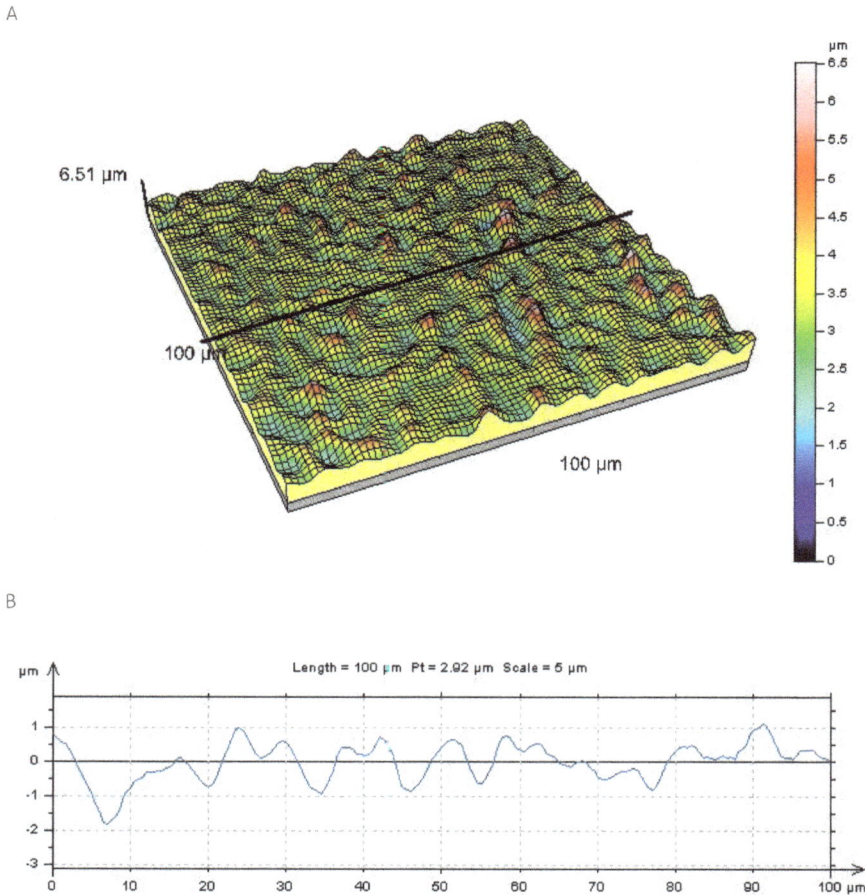

A

B

Figure 1.5 PMMA2400 surface. (A) Three-dimensional topography. (B) Two-dimensional profile (extracted at 50 µm from the edge of the square measured area).

spatial design. In the PMMA2400 profile, the wavelength $\lambda_{PMMA2400} \approx 7$-$8\ \mu m$ with some scattered irregularities can be recognized, whereas PMMA800 shows a wavelength $\lambda_{PMMA800} \approx 10$-$12\ \mu m$ and a semi regular sinusoidal superficial profile. The quantitative difference to emphasize between these two surfaces is the height of the peaks and valleys: the PMMA800 surface presents an absolute height $h_{PMMA800} \approx 2\ \mu m$, which is double the height of the peaks and valleys for PMMA2400 ($h_{PMMA2400} \approx 1\ \mu m$).

A

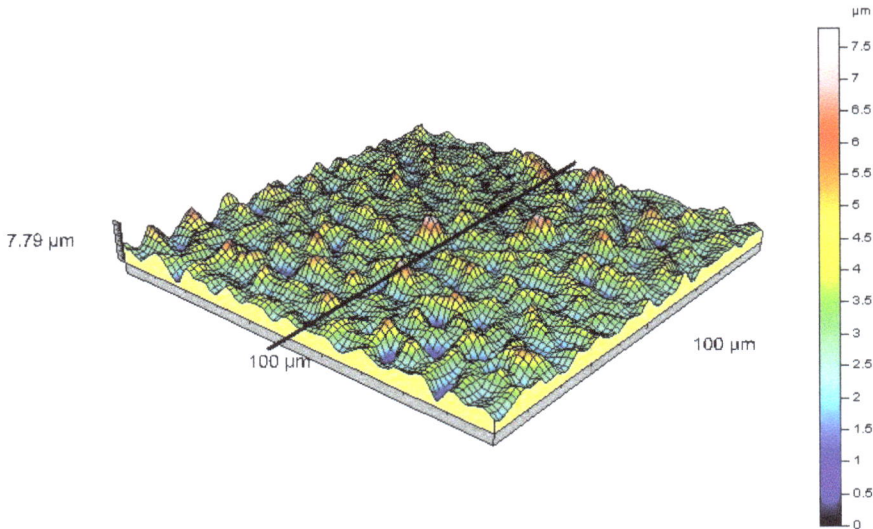

B

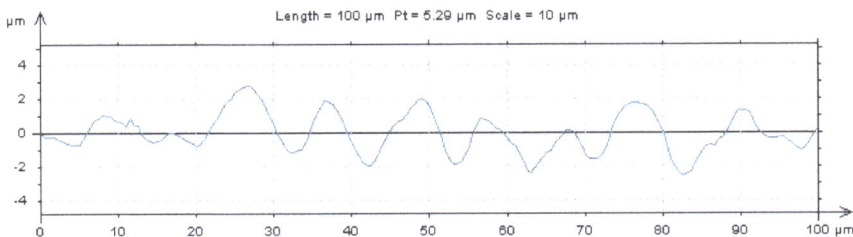

Figure 1.6 PMMA800 surface. (A) Three-dimensional topography. (B) Two-dimensional profile (extracted at 50 μm from the edge of the square measured area).

1.3. Weibull Statistics

We analysed the gecko's adhesion times using the well-known Weibull statistics. This analysis is usually applied to describe the strength and fatigue life of solids since it is based on the weakest link concept. Thus, we treat the gecko's detachment as an interfacial failure. The significant statistical correlation discovered in the presented analysis suggests the existence of a weakest link in the animal's adhesion, which is rigorously quantified by the Weibull shape and scale parameters by data fitting.

Accordingly, the distribution of failure (F) describing the cumulative probability for the gecko's detachment is expected to be:

$$F(t;m;t_0) = 1 - e^{-\left(\frac{t}{t_0}\right)^m} \tag{1}$$

where t is the measured adhesion time, m is the shape parameter (governing the standard deviation), or Weibull modulus, and t_0 is the scale parameter (governing the mean value) of the distribution of failure.

The cumulative probability $F_i(t_i)$ can be obtained experimentally as

$$F_i(t_i) = \frac{i - 1/2}{N}, \tag{2}$$

where N is the total number of measured adhesion times t_i and $t_1, \ldots, t_N$, are ranked in ascending order.

All experiments were performed at room temperature (~23 °C) and humidity (~75 %). Each set of measurements was performed on different days. The time between one measurement and the following within the same set is only the time needed to rotate the box (~14 s) in order to place the gecko again in a downward position.

1.3.1. The First Set of Experiments

We considered a female gecko (G1) adhering on inverted virgin PMMA and glass surfaces under only its own weight (~46 g). The animal was placed in its natural position on the horizontal bottom of a box (50 cm x 50 cm x 50 cm) composed of the characterized surfaces. The box was slowly rotated until the gecko reached a downward position; at that time the gecko's adhesion time was measured. We excluded any trial in which the gecko walked on the inverted surface, and the time measurement was stopped when the gecko broke loose from the inverted surface and jumped on the bottom of the box. A similar experiment was performed with a male gecko (G2, weight of ~72 g), but in this case the time was stopped at the first detaching movement of the gecko's feet. The different measurement strategies do not significantly affect the statistics of the results, confirming their robustness.

Fig.1.7 presents Weibull statistics applied to the results of the five measurements of adhesion of G1 on a virgin PMMA surface; only one set is taken during the moult (X-dots). Similarly, Fig.1.8 shows the Weibull interpretation of four sets of G1 and two of G2 (dashed lines) on the glass surface.

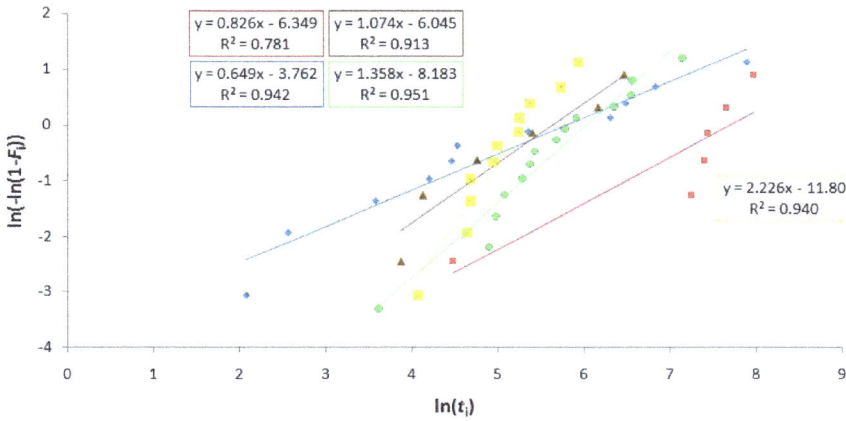

Figure 1.7 Weibull statistics applied to the four data sets for G1 and in the case of moulting (X-dots) on virgin PMMA.

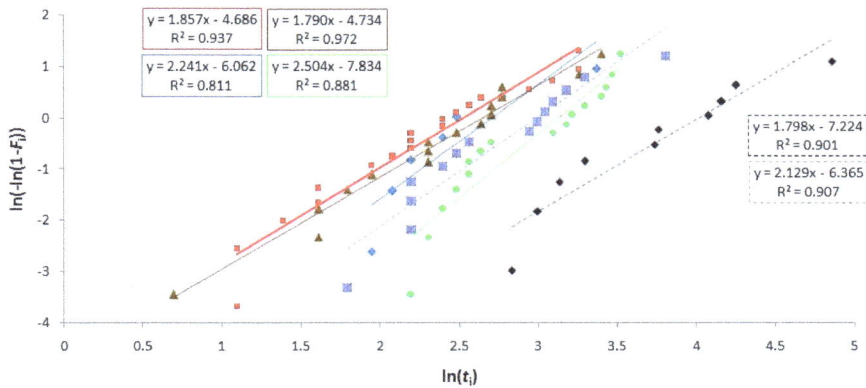

Figure 1.8 Weibull statistics applied to the four data sets for G1 and the two data sets for G2 (dashed lines) on glass.

Table 1.2 summarizes the values of the Weibull modulus, m (shape parameter), and the scale parameter, t_0, for each set on virgin PMMA (m_{PMMA} and t_{0PMMA}) and glass (m_{Glass} and t_{0Glass}).

Table 1.2 Weibull Modulus m (shape parameter) and the scale parameter $t0$ for each gecko and set, on virgin PMMA (m_{PMMA} and t_{0PMMA}) and on glass (m_{Glass} and t_{0Glass}).

		Virgin PMMA		Glass	
		Scale parameter m_{PMMA}	Shape parameter t_{0PMMA} (s)	Scale parameter m_{Glass}	Shape parameter t_{0Glass} (s)
Gecko G1	1 set	0.826	2178.6	1.857	12.5
	2 set	1.074	278.2	1.79	14.1
	3 set	0.649	329.2	2.241	15.0
	4 set	1.358	413.9	2.504	22.8
Gecko G2	1 set	\	\	1.798	55.6
	2 set	\	\	2.129	19.9
Average value		**0.977**	**800.0**	**2.053**	**23.3**

We observed the adhesive ability of G1 after moulting which significantly increased the time of adhesion by 1000 % for PMMA and 20000 % for glass, corresponding to adhesion times of hours. This again confirms that the predominant limitation of the gecko's detachment is its adhesive ability and that such an ability is limited by pollutant factors that are efficiently removed by the moult.

Considering the average values for the virgin PMMA surface, the Weibull modulus is found to be $m_{PMMA} \approx 1$ and the scale parameter t_{0PMMA} is exactly 800 s, corresponding to 13 minutes and 20 s. In each set of measurements on the PMMA surface, the correlation R^2 is high, showing coefficients of correlation $R^2 > 0.7$ in the first set and $R^2 > 0.9$ in the following three sets. For the average values for the glass surface, the Weibull modulus is found to be $m_{Glass} \approx 2$ and the scale parameter t_{0Glass} is ≈ 23 s, one order of magnitude less than the value of the scale parameter found for virgin PMMA. In each set of measurements on the glass surface, the correlation R^2 is high, showing coefficients of correlation $R^2 > 0.9$ in the first two sets using G1 and in the two sets using G2, and $R^2 > 0.8$ in the following two sets using G1.

Weibull statistics were also applied to the set of failure times (Fig.1.7) represented by the yellow dotted line set. These data were gathered during the gecko's moulting process on the virgin PMMA surface. Its capacity for adhesion is reduced to a few minutes. In this case, the Weibull modulus (m_{PMMA-M}) is found to be ~2.2 and the correlation is $R^2 = 0.94$. The scale parameter $t_{0PMMA-M}$ is ~200 s, corresponding to 3 minutes and 20 s. As the failure shape parameter m suggests, the time values have a tendency to be closer to one another. When the shape parameter m is high, very high or very low adhesion times become less probable. The probability distribution is less spread over all possible values and becomes symmetric to the scale parameter value t_0 and the time failure process becomes almost deterministic. The failure times have a magnitude of hundreds of seconds (Table 1.3, Test 1).

Table 1.3 Adhesion times on virgin PMMA (Test 1) during the moult, (Test 2) on glass surface, and (Test 3) on virgin PMMA not during the moult.

Test 1		Test 2		Test 3	
Test n°	Time (s)	Test n°	Time (s)	Test n°	Time (s)
1	59	1	9	1	8
2	104	2	10	2	13
3	108	3	11	3	36
4	108	4	12	4	67
5	142	5	13	5	87
6	148	6	13	6	93
7	190	7	14	7	212
8	192	8	15	8	550
9	216	9	22	9	660
10	310	10	24	10	936
11	380	11	25	11	2703
		12	27		
		13	30		
		14	31		
		15	32		
		16	34		

The adhesion tests on glass demonstrate a similar trend in the distribution probability function. In this case, the averaged value (m_{Glass}) emerges to be equal to 2, showing an inferior dispersion of measured adhesion time data (magnitude of tens of seconds (Table 1.3, Test 2)). The virgin PMMA adhesion tests instead present a shape parameter value (m_{PMMA}) equal to 1. The lower the parameter m, the more variable the failure time. In this case, the values are strongly variable over two orders of magnitude (Table 1.3, Test 3).

As our results confirm, on the PMMA surface the gecko shows higher failure times. Thus, we can form the hypothesis that, on PMMA surfaces during high time intervals, different causes of detachment can be introduced and be very variable, linked to the external factors of the experimental box, *e.g.* sound, light, and movement, or physiological factors of the gecko, *e.g.* hunger, cooling, disinterest, and muscular fatigue. On the other hand, on glass and virgin PMMA surfaces during moulting, the ability of the gecko to remain attached drastically decreases. The gecko realizes adhesion with several difficulties, and detachment occurs when it is unable to remain attached any longer. This is the explanation for the narrower spread of values obtained from these test measurements. As a consequence, on glass and on virgin PMMA surfaces during moulting, failure of the gecko's adhesive system certainly occurs at the instant of detachment and so the shape parameter m_{Glass}, m_{PMMA-M} and the scale parameter t_{oGlass}, $t_{oPMMA-M}$ have been correctly estimated. m_{Glass} and t_{oGlass} correspond to the real adhesive capabilities and characteristics of the Tokay gecko's foot system on glass surfaces.

According to the above-mentioned causes for detachment on PMMA surfaces, the gecko's detachment can also be linked to a limited number of variable physiological and external factors. Moreover, the test condition plans to use only two geckos. In both our experiments, we checked firm values of m_{PMMA} and t_{oPMMA} for the Tokay gecko on PMMA surfaces and the subsequent relation between the failure of the gecko's adhesive system and the surface roughness. We start from two basic concepts: two different methods used for the measurement of failure time and the observed repeatability of trials of two geckos on different and successive days. The two methods used for G1 and G2 evaluated completely and conceptually different intervals of failure times. We can form the hypothesis that our two measurement methods allowed us to exclude that the physiological factors, linked to the will and to the decisional capability of geckos, deeply influenced the results of our study. Furthermore, by repeating the experiments on different days the negative influence on results due to external factors was limited. We performed several trials during the same day and on the next days. The obtained results confirm those already obtained and permits us to suppose that estimated values for the shape parameter m_{PMMA} and the scale parameter t_{oPMMA} are correctly linked to the failure of the gecko's adhesive system.

Moreover, the greater difficulty noted for the gecko's adhesion on the glass surface could be explained through the observation of the glass profile (Fig.1.4B). It is characterized by a higher irregularity compared to the virgin PMMA profile (Fig.1.3B). The height of peaks and valleys is of the same order of magnitude compared to the virgin PMMA profile height (h_{Glass} = h_{PMMA} ≈ 0.5-1 µm), but the lateral width (w_{Glass} ≈ 2-3 µm) is almost one order of magnitude lower than that of the virgin PMMA profile (w_{PMMA} ≈ 8-9 µm). This feature suggests a considerable closeness of one peak to the next Thus, considering this superficial conformation, we can interpret our results in the following way. At the gecko toe-surface interface, the glass surface with thin, marked peaks and valleys cannot forma complementary surface to the gecko's toe. . The gecko's toe is unable to cling to the surface by nanometric contact of each singular lamella, and the gecko's adhesion ability drastically decreases. On the other hand, perhaps thanks to the presence of larger hollows and peaks, the virgin PMMA surface guarantees good adhesion despite this unfavorable situation for the two geckos before the moulting process.

The totality of above-mentioned polluting substances trapped at the foot-surface interface clearly disappears in the days that follow moulting and during which we registered an extraordinary increase in the gecko's adhesive ability.

1.3.2. The Second Set of Experiments

We have also tested machined PMMA2400/800 surfaces. Fig.1.9 presents Weibull statistics applied to the results of one set of gecko G1 on PMMA2400 and two sets of gecko G2 on both PMMA2400 and PMMA800. Table 2.4 summarizes the values of the Weibull modulus m (shape parameter) and the scale parameter t_0 for each set on PMMA2400 ($m_{PMMA2400}$ and $t_{0-PMMA2400}$) and PMMA800 surfaces ($m_{PMMA800}$ and $t_{0-PMMA800}$). On PMMA2400, the Weibull moduli for G1 and G2 are very similar the same and correspond to $m_{PMMA2400-G1} \approx m_{PMMA2400-G2} \approx 1.2$ with a statistical correlation $R^2 = 0.95$. For gecko G1, the scale parameter is $t_{0-PMMA2400-G1} \approx 1618$ s (corresponding to almost 27 minutes); the scale parameter for gecko G2 is $t_{0-PMMA2400-G2} \approx 886$ s (approximately corresponding to 15 minutes). On PMMA800, the identified Weibull modulus is $m_{PMMA800-G2} = 1.1$ and the correlation is $R^2 = 0.83$. The scale parameter of gecko G2 is $t_{0-PMMA800-G2} \approx 108$ s (corresponding approximately to 1 minute and 48 s).

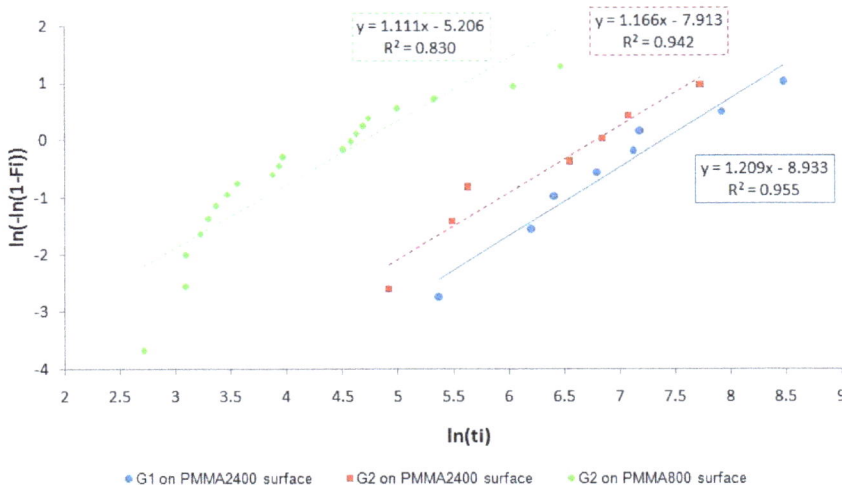

Figure 1.9 Weibull statistics applied to the data set of G1 on PMMA2400 and to the two data sets of G2 on PMMA2400 and PMMA800.

Table 1.4 Weibull Modulus m (shape parameter) and the scale parameter t_0 for each gecko and set on PMMA2400 ($m_{PMMA2400}$ and $t_{0-PMMA2400}$) and PMMA800 ($m_{PMMA800}$ and $t_{0-PMMA800}$).

		PMMA2400		PMMA800	
		Scale parameter $m_{PMMA2400}$	Shape parameter $t_{0PMMA2400}$ (s)	Scale parameter $m_{PMMA800}$	Shape parameter $t_{0PMMA800}$ (s)
Gecko G1	1 set	1.209	1617.7	\	\
Gecko G2	1 set	1.166	885.8	\	\
	2 set	\	\	1.111	108.4
Average value		**1.188**	**1251.7**	**1.111**	**108.4**

Considering the analysed virgin and machined PMMA surfaces, we have found a value of the Weibull modulus (m_{PMMA}) in the restricted range of 1-1.2, which suggests that this value is a characteristic of the PMMA/gecko system. Moreover, when comparing PMMA2400 and PMMA800, it is noteworthy that $t_{0-PMMA800}$ is one order of magnitude lower than $t_{0-PMMA2400}$ and eight times lower than $t0_{PMMA}$.

The analysis of our results and the characteristics of the studied surfaces has described an inverse relationship between the gecko's adhesive ability and the grade of roughness. Geckos show a weakening of adhesion on PMMA surfaces as roughness increases. When roughness is very high, adhesive abilities drastically decrease. These observations contradict those reported in an interesting paper by Huber (Huber, 2005b). Our hypothesis concerns the 3D complexity of PMMA surfaces and the capability of deformation and adaptability of the gecko's feet. For our focus, the *Ssk* parameter is less significant since it demonstrates a clear difference between virgin PMMA surfaces and machined ones but provides no further factors to evaluate when distinguishing different machined PMMA surfaces (PMMA2400 and PMMA800). Therefore, *Ssk* does not justify the evident decrease in adhesive ability of geckos between PMMA2400 and PMMA800 measurement tests. In the discussion of our results, the *Sdr* parameter is shown to be more significant. The roughness parameter doubles between PMMA2400 and PMMA800. The lamellae and setae on the gecko's foot can presumably adapt well to the interacting substrates, but the lamellae have shown a physiological limit. When the PMMA surface is smooth (*Sdr* < 1 %), the gecko's setae (represented in blue in Fig.1.10) and spatulae can adapt to the surface and permit van der Waals forces to act. However, the extraordinary capabilities of nanocontact hairs (spatulae) are not exploited at the top. Virgin PMMA shows that the values of the *Sa* and *Sq* parameters are lower than the spatula contact area, approximated as a circle of radius 100-200 nm (Ruibal, 1965; Dai, 2002; Haeshin, 2007; Pugno, 2007a). In this case, the spatulae cannot follow the roughness of the surface and thus cannot penetrate the characteristic valleys of virgin PMMA and adhere to the side of each individual one (Fig.1.10A). An increase in the gecko's adhesive abilities was observed for PMMA2400 with an intermediate Sdr value ($Sdr \approx 15\%$). This surface is characterized by a higher superficial complexity compared to virgin PMMA and so the real area of contact with the gecko's foot is greater. The gecko's setae and spatulae have demonstrated adaptability to the PMMA2400 surface roughness, adhering this time to the top and side of single peaks. In this way, the effective number of spatulae-surface nanocontacts increases and the gecko's adhesion increases (Fig.1.10B). Adhesion time drastically decreases on a high complexity surface ($Sdr \approx 30\%$). We suppose that, in this case, the waviness characterizing the superficial roughness ($\lambda_{PMMA800} \approx 10\text{-}12$ µm and $h_{PMMA800} \approx 2$ µm) is superior compared to the adaptability of the gecko's lamellae. As a consequence we observed a decrease in the number of setae and spatulae that interact to form a nanocontact with the surface (Fig.1.10C).

Our interpretation of the results also explains the presence of claws on the tip of each gecko toe. The claws are a fundamental additional help for geckos on surfaces with high superficial complexity where its lamellae, and thus its sub-hierarchical micro- and nano- structures (setae and spatulae), are not sufficient to guarantee secure adhesion. These surfaces must possess a level of roughness that permits claws to cling (presumably Sq of tens or hundreds of µm) and perform a secure attachment.

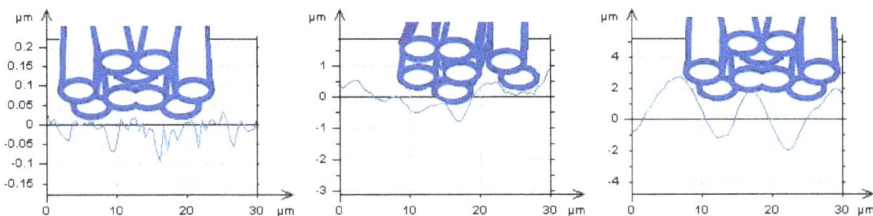

Figure 1.10 Interpretation of experimental results for adhesion tests on various roughness of PMMA surfaces. (A) Setae (represented in blue) and sub-hierarchical structures (spatulae) can adapt well on virgin PMMA; (B) The adhesion on PMMA2400 is better because of the higher number of spatulae-substrate interactions; (C) On PMMA800, only partial contact interactions are achieved.

1.4. Conclusions

In this study, we demonstrate that living geckos display adhesion times that can be described by Weibull statistics by performing three-dimensional surface topography characterizations and measurements of adhesion times. The Weibull shape (*i.e.*, modulus) and scale parameters can be used to quantitatively describe the statistics of adhesion times for different geckos (male and female), materials (glass and PMMA), and interfaces (virgin and machined PMMA surfaces).

Chapter 2

The Gecko's Optimal Adhesion on Nanorough Surfaces

Abstract

In this study, we report experimental observations of adhesion times for living Tokay geckos on inverted Poly(methyl meth-acrylate) (PMMA) surfaces. Two different geckos (male and female) and three surfaces with different root mean square (RMS) roughness (RMS = 42 nm, 618 nm and 931 nm) have been considered for a total of 72 observations. The measured data are proven to be statistically significant following Weibull statistics with correlation coefficients between 0.78 and 0.96. An unexpected result is the observation of the gecko's maximum adhesion on the surface with intermediate roughness of RMS = 618nm and waviness comparable to the seta size.

2.1. Introduction

The Tokay gecko's (*Gekko gecko*) ability to "run up and down a tree in any way, even with the head downwards" was first observed by Aristotle, almost 25 centuries ago, in his *Historia Animalium*. The pioneer study on the gecko's adhesion by Hiller (Hiller, 1968) provided Scanning Electron Microscope (SEM) pictures of the setae, showing their hierarchical ultrastructure and high density of terminal spatulae. With a very careful experiment on living geckos, his results show adhesion dependence on the surface energy of the substrate. The structure of the digital setae of lizards has been thoroughly discussed (Ruibal, 1965), but only recently the adhesive force of a single gecko foot-hair has been measured (Autumn, 2000). Like geckos, other creatures, such as beetles, flies and spiders, show comparable adhesive mechanisms and adhesive abilities resulting in an extraordinary ability to move on vertical surfaces and ceilings. A comparison between the nanostructured feet of a gecko and spider is reported in Fig.2.1.

Surface roughness strongly influences the animal's adhesion strength and ability. Its role was shown in different measurements on flies and beetles walking on surfaces with well-defined roughness (Dai, 2002; Persson, 2003; Peressadko, 2004); on the chrysomelid beetle *Gastrophysa viridula* (Gorb, 2001a), on the fly *Musca domestica* (Peressadko, 2004); on the Tokay gecko (Huber, 2007). In previous studies (Peressadko, 2004; Gorb, 2001a), a minimum

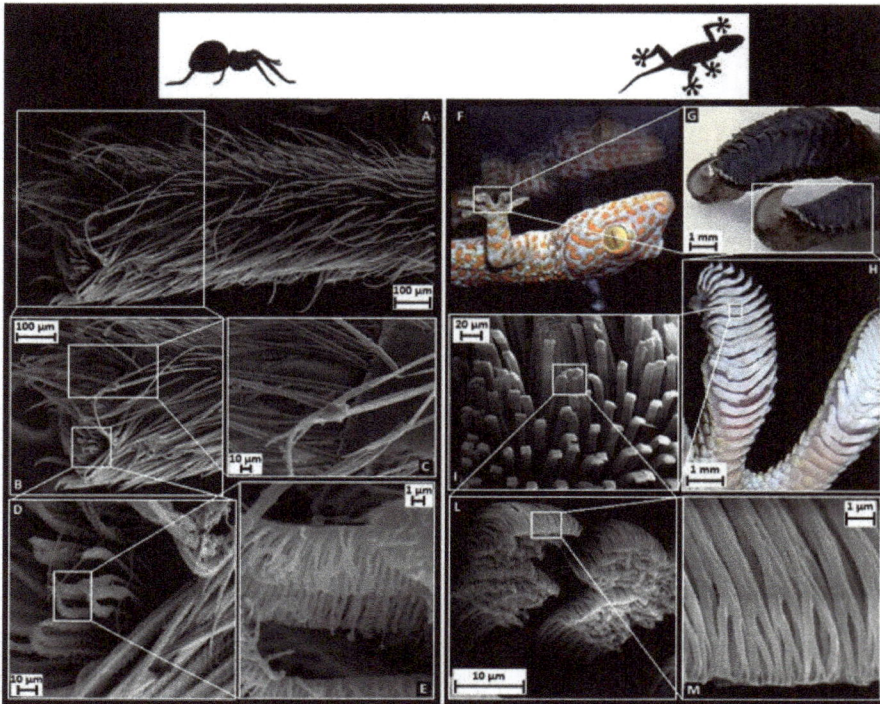

Figure 2.1 Spider and gecko feet showed by SEM (Zeiss EVO 50). The Tokay gecko (Fig.2.1F) attachment system is characterized by hierarchical hairy structures, which start with macroscopic lamellae (soft ridges ~1 mm in length, Fig.2.1H) and branches into setae (30-130 μm in length and 5-10 μm in diameter, Fig.2.1I, 2.1L (Hiller, 1968; Ruibal, 1965; Russell, 1975; Williams, 1982)). Each seta consists of 100-1000 substructures called spatulae, the contact tips (0.1-0.2 μm wide and 15-20 nm thick, Fig.2.1M (Hiller, 1968; Ruibal, 1965)) responsible for the gecko's adhesion. Terminal claws are located at the top of each singular toe (Fig.2.1G). Van der Waals and capillary forces are responsible for the resultant adhesive forces (Autumn, 2002a; Sun, 2005a), whereas the claws guarantee an efficient attachment system on surfaces with very large roughness. Similarly, an analogous ultrastructure is found in spiders (*e.g. Evarcha arcuata* (Kesel, 2003)). Thus, in addition to the tarsal claws, which are present on the tarsus of all spiders (Fig.2.1C), adhesive hairs can be distinguished in many species (Fig.2.1D, 2.1E). Like for insects, these adhesive hairs are specialised structures that are not restricted only to one particular area of the leg, but may be found either distributed over the entire tarsus - as for lycosid spiders - or concentrated on the pretarsus as a tuft (scopula) situated ventral to the claws (Fig.2.1A, 2.1B) - as in the jumping spider *Evarcha arcuata* (Kesel, 2003).

of the adhesive/frictional force, spanning surface roughness from 0.3 µm to 3 µm, is reported. Experiments on the reptile Tokay gecko (Huber, 2003) show a minimum in the adhesive force of a single spatula at an intermediate root mean square (RMS) surface roughness around 100-300 nm and a monotonic increase of adhesion times for living geckos by increasing the RMS from 90 to 3000 nm. There are several observations and models in the literature, starting with the pioneer paper by Fuller and Tabor (Fuller, 1975), in which roughness is seen to decrease adhesion monotonically. However, there is also experimental evidence, starting with the pioneer paper (Briggs, 1977) that suggests that roughness need not always reduce adhesion. For example, in the framework of a reversible model (Persson, 2001; Persson, 2002), it has been shown that for certain ranges of roughness parameters, it is possible for the effective surface energy to first increase with roughness amplitude and then eventually decrease. Including irreversible processes, due to mechanical instabilities, it has been demonstrated (Guduru, 2007) that the pull-out force must increase by increasing the surface wave amplitude. Our results suggest that roughness alone is not sufficient to describe the three-dimensional topology of a complex surface, and additional parameters have to be considered for formulating a well-posed problem.

Accordingly, we have machined and characterized three different Poly(methyl meth-acrylate) (PMMA) surfaces (PMMA 1, 2, 3; surface energy of ~41 mN/m) with a full set of roughness parameters as reported in Table 2.1 (see paragraph 1.2 for a detailed explanation of the extracted classical roughness parameters (Sa, Sq, Sp, Sv, Sz, Ssk, Sdr)).

2.2. Materials and Methods

Two different Tokay geckos, female (G1, weight of ~46 g) and male (G2, weight of ~72 g), have been considered. The gecko was placed in its natural position on the bottom of a box (50 cm x 50 cm x 50 cm). Then, box was slowly rotated until the gecko reached a natural downwards position, at which time the gecko's adhesion time was measured. We excluded any trial in which the gecko walked on the inverted surface. The time measurement was stopped when gecko broke loose from the inverted surface and fell on the bottom of the box (for G1) or at the first detachment movement of the gecko's foot (for G2). The time between one measurement and the next, pertaining to the same set, was only the time needed to rotate the box (~14 s) and replace the gecko on the upper inverted surface. The experiments were performed at room temperature (~22 °C) and humidity (~75 %). The measured adhesion times are summarized in Table 2.2 and confirmed to be statistically significant by applying Weibull statistics (Fig.2.2).

2.3. Results and Conclusions

We have observed a maximum in the gecko's adhesion times on PMMA 2, which has an intermediate roughness of RMS = 618 nm. An oversimplified explanation could be the following. For PMMA 1 (Sq_{-PMMA1} = 42 nm, Sdr_{-PMMA1} < 1 % and waviness of λ_{PMMA1} ≈ 3-4 µm, h_{PMMA1} ≈ 0.1 µm), the gecko's setae (diameter of ~10 µm, represented in blue in Fig.2.3, and must not be confused with the terminal nearly two-dimensional spatulae) cannot penetrate in the characteristic smaller valleys and adhere to the side of each single one (Fig.2.3A); thus the gecko cannot optimally adapt to the surface roughness. For PMMA 2 (Sq_{-PMMA2} = 618 nm, Sdr_{-PMMA2} ≈ 15 % and λ_{PMMA2} ≈ 7-8 µm h_{PMMA2} ≈ 1 µm), the gecko's setae are better able to adapt to the roughness, adhering this time on the top of and on the side of a single asperity. In this way, the effective number of setae in contact with the surface increases and, as a direct consequence, the adhesive ability of the gecko also increases (Fig.2.3B). On PMMA 3 (Sq_{-PMMA3} = 931 nm, Sdr_{-PMMA3} ≈ 30 % and λ_{PMMA3} ≈ 10-12 µm, h_{PMMA3} ≈ 2 µm), the waviness characterizing the roughness is larger than the size of the setae. Thus, a decrease in the number of setae in contact is expected (Fig.2.3C), and on PMMA 2, an adhesion increment of about 45 % is observed. According to a previous study (Briggs, 1977), an increment of 40 %, close to our observation, is expected for an adhesion parameter α equal to 1/3. Such a parameter was introduced as the key parameter governing adhesion (Fuller, 1975):

$$\alpha = \frac{4\sigma}{3}\left(\frac{4E}{3\pi\sqrt{\beta\gamma}}\right)^{2/3} \tag{1}$$

where σ is the standard deviation of the asperity height distribution (assumed to be Gaussian), β is the mean radius of curvature of the asperity, γ is the surface energy, and E is the Young modulus of the soft solid (the gecko's foot). Even if the value of E for the entire foot cannot be simply defined as a consequence of its non-compact structure, considering it to be of the order of 10 MPa (thus much smaller than that of the keratin material), with γ ≈ 0.05 N/m (Autumn, 2000), σ ≈ Sq, β ≈ λ would correspond to values of α close to 0.5.

The reported maximum adhesion was not observed by Huber (Huber, 2003). It is noteworthy that their tested polished surfaces were of five different types, with a nominal asperity size of 0.3 µm, 1 µm, 3 µm, 9 µm, and 12 µm, which correspond to RMS values of 90 nm, 238 nm, 1157 nm, 2454 nm, and 3060 nm, respectively. The sliding of geckos on polishing paper with a RMS value of 90 nm for slopes larger than 135° was observed (Huber, 2003). On a rougher substrate, with a RMS value of 238 nm, two individual geckos were able to cling to the ceiling, but the foot-surface contact had to be continuously renewed because geckos' toes tended to slowly slide off the substrate. Finally, on the remaining rougher substrates tested, animals were able to adhere stably to the ceiling for more than 5 minutes.

These different observations (assuming that the influences of claws and moult were also minimized (Huber, 2003)) suggest that the RMS parameter is not sufficient alone to describe the aspects of surface roughness (Lepore, 2008; Lepore, 2010; Pugno, 2008a; Pugno 2008b). The use of a "complete" set of roughness parameters could help in better understanding the animal's adhesion.

Table 2.1 Roughness parameters for the three different PMMA (1, 2, 3) surfaces.

	PMMA1	PMMA2	PMMA3
S_a (μm)	0.033 ± 0.0034	0.481 ± 0.0216	0.731 ± 0.0365
S_q (μm)	0.042 ± 0.0038	0.618 ± 0.0180	0.934 ± 0.0382
S_p (μm)	0.252 ± 0.0562	2.993 ± 0.1845	4.620 ± 0.8550
S_v (μm)	0.277 ± 0.1055	2.837 ± 0.5105	3.753 ± 0.5445
S_{sk}	-0.122 ± 0.1103	0.171 ± 0.1217	0.192 ± 0.1511
S_z (μm)	0.432 ± 0.1082	4.847 ± 0.2223	6.977 ± 0.2294
S_{dr} (%)	0.490 ± 0.0214	15.100 ± 1.6093	28.367 ± 2.2546

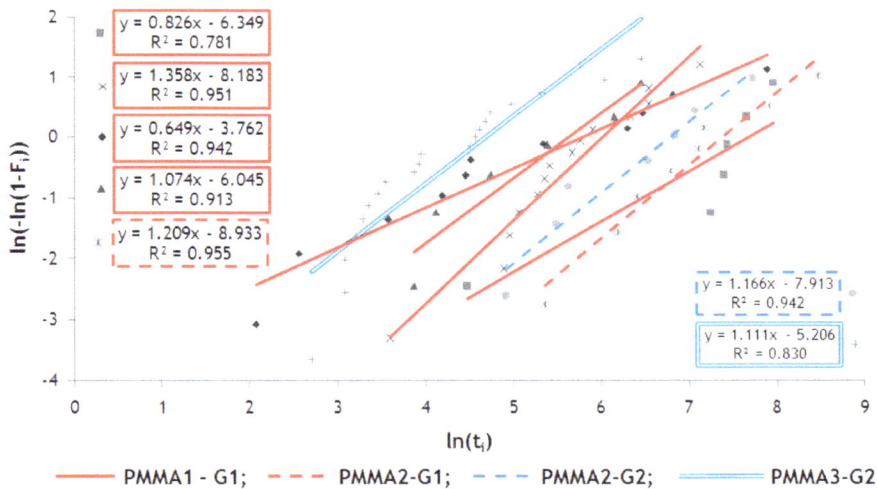

Figure 2.2 Weibull statistics (F is the cumulative probability of detachment/failure and *ti* is the measured adhesion time) applied to the measured adhesion times on PMMA surfaces. PMMA 1 (red lines, for which we made 4 sets of measurements on four different days with gecko G1), PMMA 2 (dotted lines, for which we made 2 sets of measurements on two different days, one with gecko G1 (red) and one with gecko G2 (blue)), and PMMA 3 (blue double-line, for which we made the measurements in a single day with gecko G2).

Table 2.2 The gecko's adhesion times on PMMA 1, 2, 3 surfaces. Note that, as an index of the gecko's adhesive ability, we use the Weibull scale parameter t_0 (in seconds) for the distribution of the detachment/failure F (closely related to its mean value).

Test No.	PMMA 1	PMMA 2	PMMA 3
1	8	137	15
2	13	215	22
3	36	243	22
4	37	280	25
5	48	498	27
6	62	610	29
7	67	699	32
8	87	900	35
9	88	945	48
10	93	1194	51
11	116	1239	53
12	134	1320	91
13	145	2275	97
14	160	2740	102
15	197	4800	109
16	212		114
17	215		148
18	221		207
19	228		424
20	292		645
21	323		
22	369		
23	474		
24	550		
25	568		
26	642		
27	660		
28	700		
29	707		
30	936		
31	1268		
32	1412		
33	1648		
34	1699		
35	2123		
36	2703		
37	2899		
Scale Parameter t_0 (s)	800	1251.7	108.4
Sq (µm)	0.042 ± 0.0038	0.618 ± 0.0180	0.934 ± 0.0382

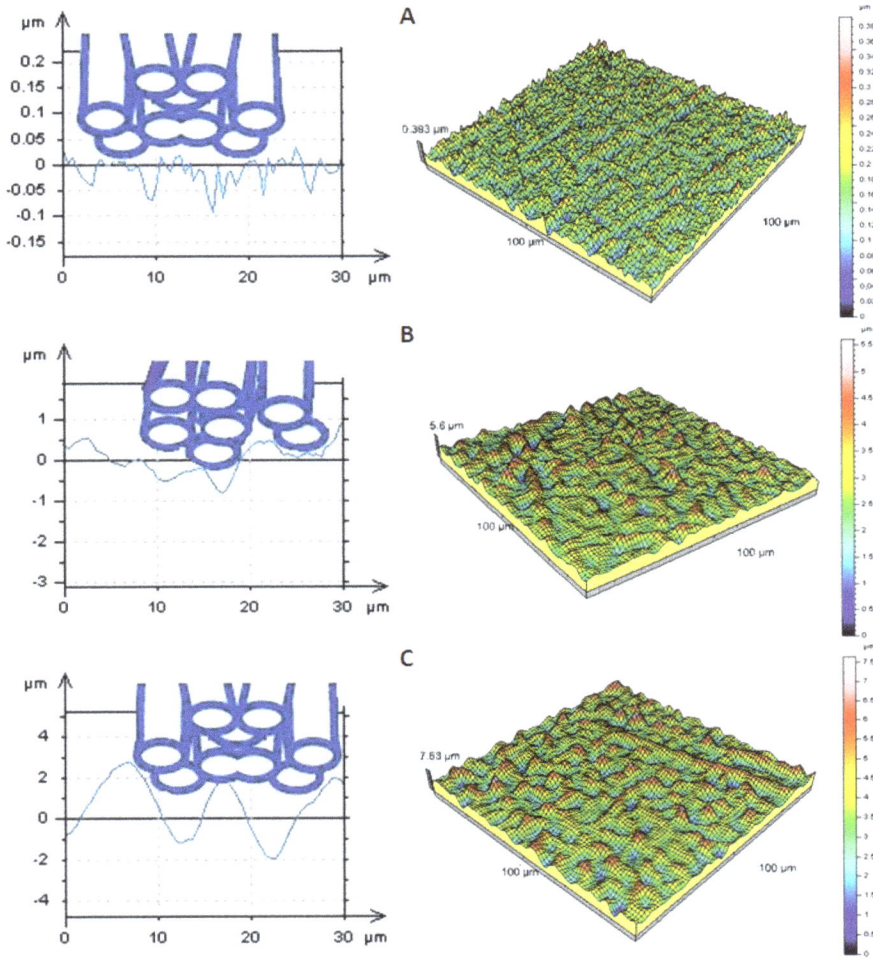

Figure 2.3 A simple interpretation of our experimental results on the adhesion tests for living geckos on PMMA surfaces with different roughness. (A) Setae cannot adapt well on PMMA 1; (B) On PMMA 2, the adhesion is enhanced because of the higher compatibility in size between setae and roughness; (C) On PMMA 3 only partial contact is achieved. On the right, we report the analysed three-dimensional roughness profiles for all three investigated surfaces (from the top: PMMA 1, 2 and 3).

Chapter 3

Normal Adhesive Force-Displacement Curves of Living Tokay Geckos

Abstract

In this study, we report experimental measurements for the normal adhesive force *versus* body displacement for living tokay geckos (*Gekko gecko*) adhering to Poly(methyl meth-acrylate) (PMMA) and glass surfaces. We have measured the normal force needed to reach the gecko's detachment. Atomic force and scanning electron microscopy are used to characterize surface and foot topologies. The measured safety factors (maximum adhesive force divided by the body weight) are 10.23 on PMMA surfaces and 9.13 on glass surfaces. We have observed minor but reversible damage to the gecko feet caused by our tests, as well as self-renewal of adhesive abilities after moulting.

3.1. Introduction

The ability of a gecko to remain stuck motionless to a vertical surface or even to a ceiling seems to defy gravity. In the 4th century B.C., geckos were been observed to "run up and down a tree in any way, even with their head downwards" by Aristotle (Aristotle, 343). Scientific researchers have focused their attention on the gecko's adhesive foot architecture, adhesive abilities, and related mechanisms (Irschick, 1996; Cartier, 1872; Haase, 1900; Gadow, 1902; Weitlaner, 1902; Schmidt, 1904; Hora, 1923; Dellit, 1934; Mahendra, 1941; Maderson, 1964; Ruibal, 1965; Hiller, 1968; Hiller, 1969; Gennaro, 1969; Williams, 1982; Stork, 1980; Liang, 2000; Autumn, 2000; Autumn, 2002c). Scanning electron microscopy (SEM) has created new opportunities to observe under the length-scale limitations imposed by the wavelength of visible light and to study the sub-micrometric hierarchical architecture of gecko toes.

The Tokay gecko (*Gekko gecko*) is the second largest Gekkonid lizard species (1050 species in the world), attaining lengths of 0.3-0.4 m and 0.2-0.3 m for males and females, respectively. The weight of an adult gecko ranges from ~30 to ~300 g (Tinkle, 1992). A previously study on Tokay geckos (Irschick, 1996) reveals a strong shear adhesive force of ~20 N when placed with front feet contacting a nearly vertical (85°) acetate sheet attached to a stiff PMMA plate. Consequently, assuming the gecko's weight is ~100 g, a shear safety factor (sSF) can be estimated to be approximately 40. This sSF is comparable with that of the *Hemisphaerota cyanea* beetle (sSF ~ 60; measured for a force applied perpendicularly to a vertical attachment surface generated either electronically or by hanging weights (Eisner, 2000)); of the *Chrysolina Polita* leaf beetle (sSF ~ 50; attached to a force transducer (Stork, 1980)); lower than that of the jumping spider *Evarcha arcuata* (sSF ~ 160; theoretically extracted *via* atomic force microscopy (AFM) analysis (Kesel, 2003)) and of *Crematogaster* cocktail ants (sSF~146; measured using a centrifuge technique (Federle, 2000)). Thus, not only for insects and spiders (Irschick, 1996; Stork, 1980; Eisner, 2000; Kesel, 2003; Federle, 2000; Walker, 1993), but also for geckos, several studies have been performed with the aim of quantifying the maximum adhesive force by direct *in vivo* (Irschick, 1996; Autumn, 2000; Pugno, 2007a; Autumn, 2002b; Autumn, 2002c; Huber, 2005a; Huber, 2005b; Huber, 2007; Autumn, 2006b; Autumn, 2006c) or *in vitro* measurements (Autumn, 2000; Pugno, 2007a; Autumn, 2002b; Autumn, 2002c; Huber, 2005b; Huber, 2007).

In this study, we report measurements of the normal adhesive force *versus* body displacement of living Tokay geckos up to surface detachment. We are also interested in comparing the effects of surface roughness on the gecko's maximum normal safety factor. The influence of damage to the gecko's feet caused by our experiment, on adhesive ability is also discussed. The surface topography of PMMA and glass was analysed by AFM, and SEM was employed to characterize the hierarchical architecture of the gecko's feet.

3.2. Materials and Methods

3.2.1. Gecko's Feet Architecture

Considering Figs.3.1-3.3, investigation of the hierarchical structure of the gecko's toes were performed using SEM (Zeiss EVO 50) equipped with a lanthanum hexaboride cathode and FESEM (Zeiss SUPRA 40) equipped with a field emission tungsten cathode. For SEM analysis, three frozen and formaldehyde-fixed toe samples retrieved from two geckos that had died naturally were unfrozen at room temperature, 5h-dehydrated with ethanol incremenetally increasing every hour (10 %, 30 %, 50 %, 70 %, 100 %). Samples were fixed to aluminium stubs with double-sided adhesive carbon conductive tape (Agar Scientific), 30-min air-dried and gold-coated (approx. 40 nm) in a SCD 050 sputter coater (BalTec). For FESEM analysis, sample preparation with the same method, but the samples were chrome-coated (approx. 20 nm).

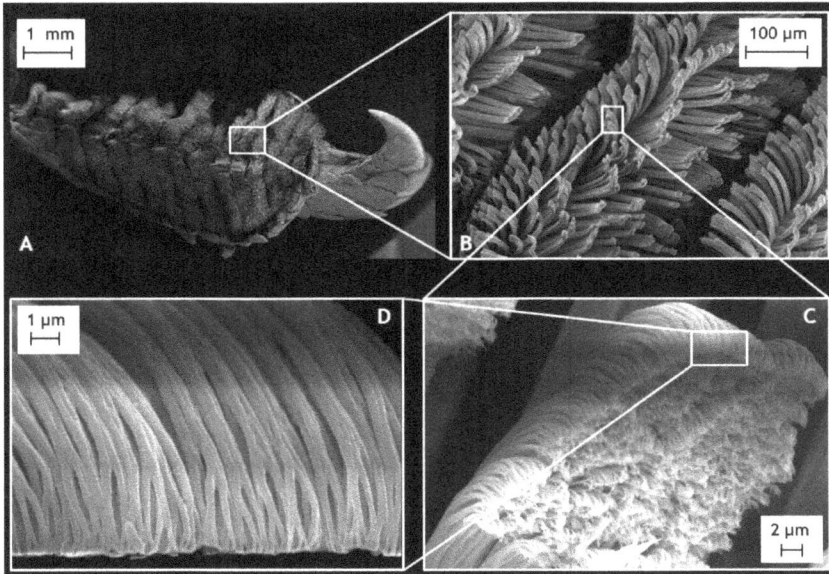

Figure 3.1 The Tokay gecko's adhesive system was observed by FESEM (Zeiss SUPRA 40) (A, B) and by SEM (Zeiss EVO 50) (C, D). (A) Toe and FESEM micrograph of the setae (B). SEM micrograph of the setae (C) A nanoscale array of hundreds of spatulae (D).

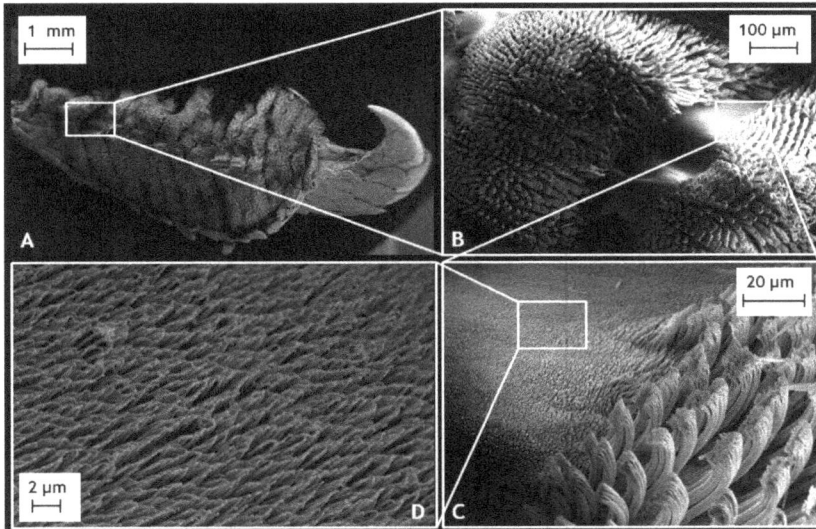

Figure 3.2 The Tokay gecko's adhesive system was observed by FESEM (Zeiss SUPRA 40). (A) The Tokay gecko's toe. (B, C) The connection area between adjacent lamellae, localized perpendicular to the longitudinal axis of each digit, is covered by nanostructured hairy units; (D) at high magnification.

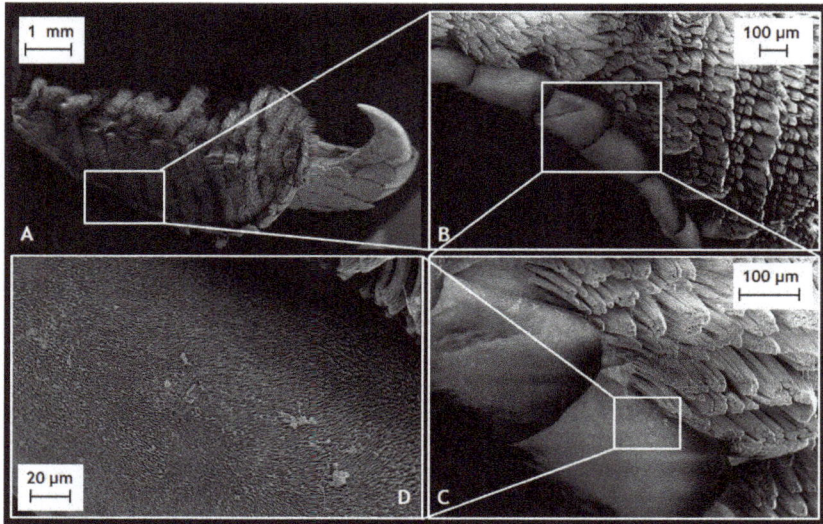

Figure 3.3 The Tokay gecko's adhesive system was observed by FESEM (Zeiss SUPRA 40). (A) The Tokay gecko's toe. (B, C) The edge of the gecko's toe is covered by nanostructured hairy units; (D) at high magnification.

3.2.2. PMMA and Glass Surface Characterization

The surface roughness of PMMA and glass was nanocharacterized by AFM Perception (Assing, Rome, Italy) using the contact mode with a silicon nitride tip. A surface area of 10 μm × 10 μm for each material was evaluated with a final resolution of 200 points/profile

3.2.3. The Gecko's Normal Adhesive Force *versus* Displacement Curves

We used a single male adult Tokay gecko (authorized by Ministerial Decree n° 73/2010-B). The gecko was maintained in a terrarium at ~28 °C. The temperature of the experimental room in which the force-displacement measurements were performed, was ~22 °C. The gecko was fed moths and water *ad libitum* and crickets once per week. The animal did not show any particular discomfort being manipulated, segregated in the test box, and bound with adherent elastic cloth bandaging.

Force-displacement measurements were conducted as follows. The gecko was prepared and placed in a PMMA-Glass (Vetronova, Varese, Italy) box 10 minutes before each set of tests. The gecko was removed from its terrarium and fixed to adherent cloth bandaging; a metallic hook was inserted within the bandage on the gecko's back. The gecko was then connected, by means of a plastic wire tied to the metallic hook, to the measurement platform and placed gently on the bottom of the measurement box (Fig.3.4). The force-displacement measurement platform was built outside the box (Fig.3.5). We applied force using an incrementally increasing mass (16, 48, 98, 148, 198, 273, 348, 423, 498, 573, 648, and 723 g). The displacements from the point of applied force on the gecko's body were recorded during the test. The measured displacement corresponds to the stretching of the front and rear legs of gecko without slipping of its feet.

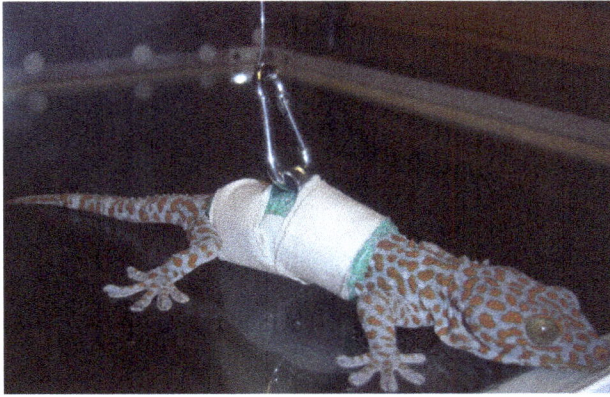

Figure 3.4 The experimental Tokay gecko with adherent elastic cloth bandaging and metallic connection hook on the measurement platform.

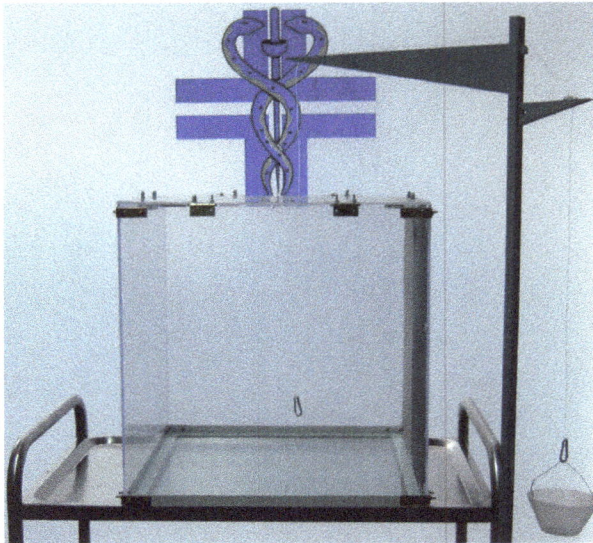

Figure 3.5 Force-displacement measurement platform.

The increase of hanging weight was conducted as follows. After the initial application of 16 g, we waited 10 seconds to record a stabilized value of the gecko's displacement on a millimetric scale. We continued similarly with the next applied weights up to 198 g. For larger weights, we allowed a relaxing time of about 15 s after each application to avoid muscular fatigue. When detachment occurred, the gecko was pulled upwards but immediately reached a secure point, approximately 42 cm from the top of the box, then was slowly taken back to the bottom. Each force-displacement curve was obtained in ~3 minutes. During a single test, the only allowed action was the renewal of foot contact and hyperextension (Irschick, 1996).

We have accordingly measured normal adhesive force-displacement curves for a gecko adhering to the interior surface of a box (50 cm x 50 cm x 50 cm). One wall of the box was glass and the other walls were PMMA. We performed fifteen tests on PMMA and three tests on glass after the first moult process and three tests on PMMA and four tests on glass after the next moult.

The first test-day, 50 days following the first moult, we performed only one force-displacement curve on both PMMA and glass (blue line, Figs.3.6 and 3.7, respectively). The second day of tests was 62 days after the first moult, during

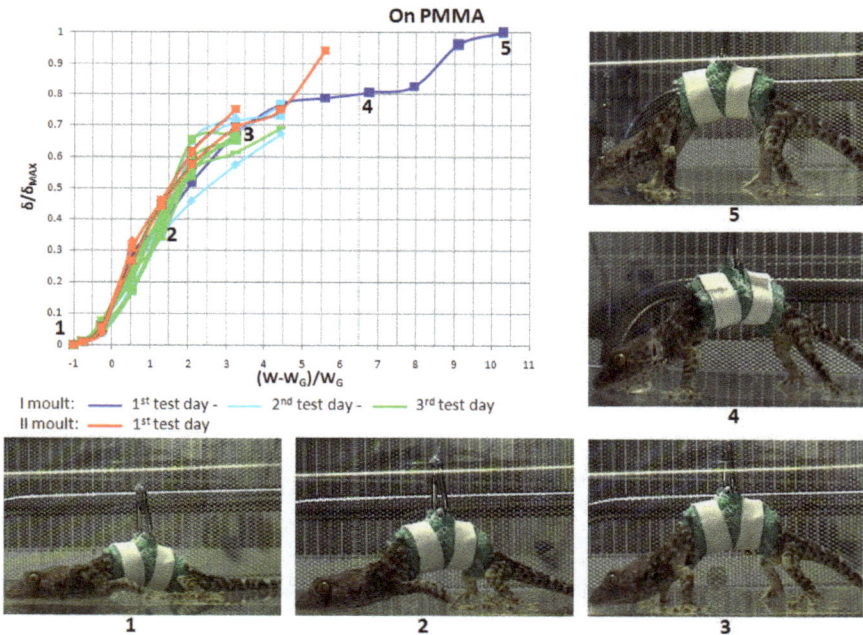

Figure 3.6 Normal adhesive force-displacement curves on PMMA surfaces after the first and second moults. Snapshots show five specific instants of the gecko's displacement at 0, 148, 273, 423, and 723 g of hanging weight (W is the applied weight, W_G is the gecko's weight, δ is the gecko's displacement, and δ_{MAX} is the gecko's maximum displacement).

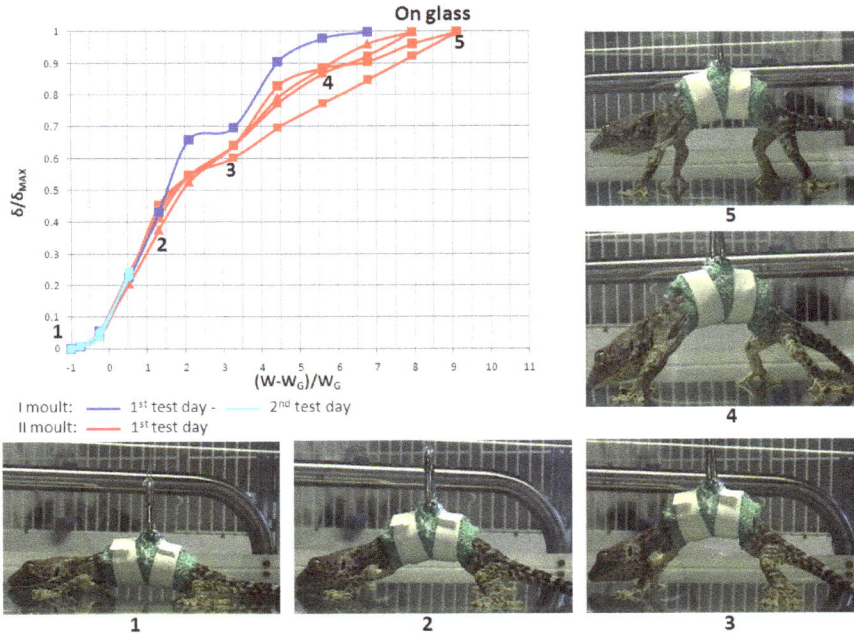

Figure 3.7 Normal adhesive force-displacement curves on glass after the first and second moults. Snapshots show five specific instants of the gecko's displacement at 0, 148, 348, 423, and 648 g of hung weight (W is the applied weight, W_G is the gecko's weight, δ is the gecko's displacement, and δ_{MAX} is the gecko's maximum displacement).

which we performed four tests on PMMA (cyan line, Fig.3.6) and two tests on glass (cyan line, Fig.3.7). The third testing day was the next day during which we performed ten tests on PMMA (green line, Fig.3.6). 7 days after the second moult, experiments were conducted in only one day, due to the damage imposed by the first day of tests on the gecko's feet. Four tests (red line, Fig.3.7) were performed on glass followed by three force-displacement curves on PMMA (red line, Fig.3.6).

3.3. Results

3.3.1. The Gecko's Feet Architecture

The Tokay gecko foot consists of five digits (Fig.3.1A) covered with macroscopic hairy structures called lamellae (~0.5-3 mm in width and 200-500 μm in length, Fig.3.1B). These lamellae are organized in a series of multi-arrays localized perpendicular to the longitudinal axis of each digit;

the lamellae are separated one from another. Nanostructured hairy units (~2-5 μm in length and ~200 nm in diameter; Figs.3.1C, 3.1D) have been identified on connection areas between adjacent lamellae (Fig.3.1C) and on the edge of each single digit (Figs.3.2B, 3.2C, 3.2D). Each lamella is covered with several thousand setae (10-130 μm in length and 3-10 μm in diameter, with a density of ~0.014 setae/μm² (Ruibal, 1965; Schleich, 1986), Figs.3.1B, 3.1C), each of which contain at their tips hierarchical substructures called spatulae (0.1-0.2 μm wide and 15-20 nm thick, Fig.3.3D). Terminal claws are located at the top of each single toe (~500 μm in diameter and ~1 mm in length, Fig.3.1A) and guarantee a secure mechanical interlocking on surfaces with high roughness, *i.e.* where the diameter of the gecko's claw tip is smaller than the roughness (Lepore, 2008; Pugno, 2008a; Pugno, 2008b; Pugno, 2008c; Varenberg, 2010; Lepore, 2010).

3.3.2. PMMA and Glass Surface Characterization

Table 3.1 summarizes roughness parameters of the studied PMMA and glass surfaces. The PMMA (Fig.3.8) and glass (Fig.3.9) surfaces have different roughness, and isolated bubbles of diameter ~1 μm are visible on the glass surface.

Table 3.1 Roughness parameters of the studied PMMA and glass surfaces.

	PMMA	Glass
R_a (nm)	3.81 ± 0.085	0.80 ± 0.214
R_q (nm)	5.88 ± 0.778	1.41 ± 0.796
R_v (nm)	52.74 ± 14.938	16.88 ± 13.895
R_p (nm)	90.06 ± 28.736	21.61 ± 16.943
S_{sk}	1.41 ± 0.997	0.79 ± 0.461
S_{dr} (%)	0.60 ± 0.046	0.02 ± 0.007

3.3.3. The Gecko's Normal Adhesive Force *versus* Displacement curves

During the first test day, a maximum SF for the PMMA surface $\lambda_{PMMA(I)-1Day}$ = 10.23 was observed. To compute this value, we consider the animal's weight (64 g) and the final hung weight of 723 g (Fig.3.6, snapshot 5). In the second test-day, the

Figure 3.8 AFM characterization of the PMMA surface.

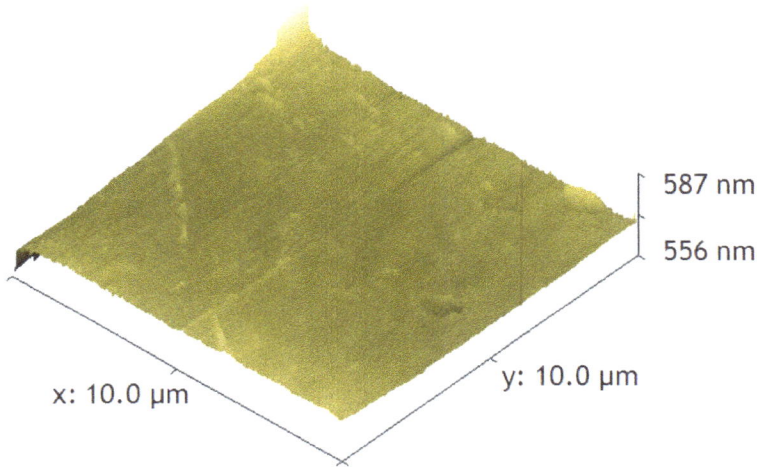

Figure 3.9 AFM characterization of the glass surface.

gecko reached an average SF reduced by 60% ($\lambda_{PMMA(I)-2Day} \approx 4.1$) in comparison with the maximum value. Finally, a minimum value for the SF equal to $\lambda_{PMMA(I)-3Day}$ ≈ 2.1 was observed during the third test day. Analogously, on the glass surface, the final hung weight of 498 g and the gecko's same weight correspond to a maximum SF $\lambda_{Glass(I)-1Day} \approx 6.8$ in the first test day. In the second test day, it was reduced to less than 1 ($\lambda_{Glass(I)-2Day} \approx 0.5$). In the first test day after the second moult, the final maximum hung weight of 648 g and the gecko's same weight correspond to a maximum SF $\lambda_{Glass(II)-1Day} = 9.13$ (Fig.3.7, snapshot 5). After four tests on the glass surface, three tests were performed on the PMMA surface, reaching an SF that gradually decreased starting from $\lambda_{PMMA(II)-1Day} \approx 5.6$ to the final minimum $\lambda_{PMMA(II)-1Day} \approx 0.5$. In summary, the final maximum SF is found to be

$\lambda_{PMMA} = \lambda_{PMMA(I)\text{-}1Day} = 10.23$ on the PMMA surface and $\lambda Glass = \lambda_{Glass(II)\text{-}1Day} = 9.13$ on the glass surface.

3.4. Discussion

Figs.3.6 and 3.7 report the measured force-displacement curves and five snapshots of the gecko's specific configurations on PMMA and glass, respectively. The obtained force-displacement curves are condensed to show overall trends. In the first test day after the first moult, the gecko's maximum SF for the PMMA surface was observed, whereas in the first test day after the second moult, the maximum SF for the glass surface was observed. The SF of ~10 that we measured for Tokay geckos is consistent with previously reported observations. In particular, a previous study (Irschick, 1996) measured the shear adhesive force by placing geckos with front feet contacting a nearly vertical acetate sheet (85°) and then slowly pulling in a downward direction. Our experimental set instead permitts us to evaluate the normal force required to detach the Tokay gecko from a horizontal surface (PMMA and glass). Thus, the maximum shear force can be estimated to be ~40 N for the living Tokay gecko (Irschick, 1996) for maximum normal adhesive force of 7.1 N on PMMA and 6.4 N on glass.

Considering a setae density of 14.000 setae/mm² (Autumn, 2002b; Huber, 2005b; Schleich, 1986) and the gecko's total pad area of 450 mm², a shear adhesive force of ~40 N (Irschick. 1996) and a normal adhesive force of ~6.7 N, imply for a single seta a shear adhesive force of 6.2 µN (Autumn, 2002a) and a normal adhesive force of 1.1 µN. These top-down computations are underestimated due to the unavoidable presence of defects at the macroscale of the pads. Indeed, the maximum shear adhesive force for a single seta was directly measured as ~200 µN (Autumn, 2000; Autumn, 2002a; Autumn, 2006c), leading to a theoretical shear adhesive force for the gecko of 1250 N (Autumn, 2002a); similarly, the maximum normal adhesive force for a single seta is ~40 µN (Autumn, 2002b; Autumn, 2006c) leading to a theoretical normal adhesive force for the gecko of 250 N (Autumn, 2002a). At the size of the spatulae, a normal adhesive force of ~10 nN has been determined (Huber, 2005a; Huber, 2007; Huber, 2005b), leading to a final adhesive force for the gecko of 65 N (if we assume that the gecko has 6.5 billion spatulae (Irschick, 1996; Autumn, 2002a, Schleich, 1986)).

From the results at different characteristic sizes, we can conclude that the force estimated at the macroscale (*i.e.* of the whole gecko) leads to an underestimation of nearly 32 times the microscale (setae) shear adhesive force and of nearly 36 times the microscale normal adhesive force; thus 'smaller is stronger' (Buehler, 2006; Keten, 2010a). Similarly, at the nanoscale (spatulae), the normal adhesive strength is nearly 10 times that at the macro-scale (Gorb, 2001b). As a consequence of the presence of defects (Pugno, 2007a; Autumn, 2006b) at the level of the entire body, a normal safety factor of ~10 is expected for secure attachment and easy detachment, as we have measured.

In summary, the shear adhesive force is equal to ~200 µN (Autumn, 2000; Autumn, 2002a; Autumn, 2006b) for a single seta and ~40 N (Irschick, 1996) for the whole gecko, whereas the normal adhesive force is equal to ~10 nN (Huber, 2005a; Huber, 2007; Huber , 2005b) for a single spatula, ~40 µN (Autumn, 2002b; Autumn, 2006b) for a single seta, and the shear adhesive force determined in this study ~7.1 (~6.4) N on PMMA (glass) for the whole gecko. Thus, our measurement of the normal adhesive force for the whole gecko contributes to the characterization of the functionality of the hierarchical adhesive system for the Tokay gecko (Buehler, 2010) and confirms the ratio of 5:1 between the shear and normal adhesive forces for the whole animal observed by Autumn (Autumn, 2006c) for a different climbing gecko (*Hemidactylus garnotii*, ~2 g of body mass). Interestingly, for Tokay gecko, a ratio of 5:1 for the shear to normal adhesive forces is verified both at macro and micro scales.

In addition, we observe the self-renewal of the gecko's adhesive system after moulting and a negative effect from the previously executed experimental tests which lead to a reduction of the maximum adhesive force.

3.4.1. Feet Damage

During the first test day after the second moult, we observed evident foot damage. As mentioned above, the four tests performed on the glass surface followed by the three tests on the PMMA surface indicate that the gecko's attachment drastically decreases from one test to the next. In particular, a decrement of the SF on the PMMA surface corresponding to 40% from the first to the second test and to ~85 % from the second to the third test was observed. After these three tests, the gecko could no longer stay attached with its hind feet. Fig.3.10 shows the negative effects of the seven consecutive tests photographed one day after the first test day subsequent to the second moult. A diffused inflammation of the gecko toes and the presence of a small thin wound located on the gecko's skin between one toe and the next were observed.

Regarding self-renewal of the gecko's adhesive system and abilities after moulting, an increase of the gecko's SF was measured from $\lambda_{Glass(I)-2Day}$ ≈ 0.5 before the second moult to $\lambda_{Glass(II)-1Day}$ = 9.13 after the second moult. The SF increase is also appreciable on the PMMA surface: an SF $\lambda_{PMMA(I)-3Day}$ ≈ 2.1 was measured before the second moult and increased to $\lambda_{PMMA(II)-1Day}$ ≈ 5.6 after the second moult.

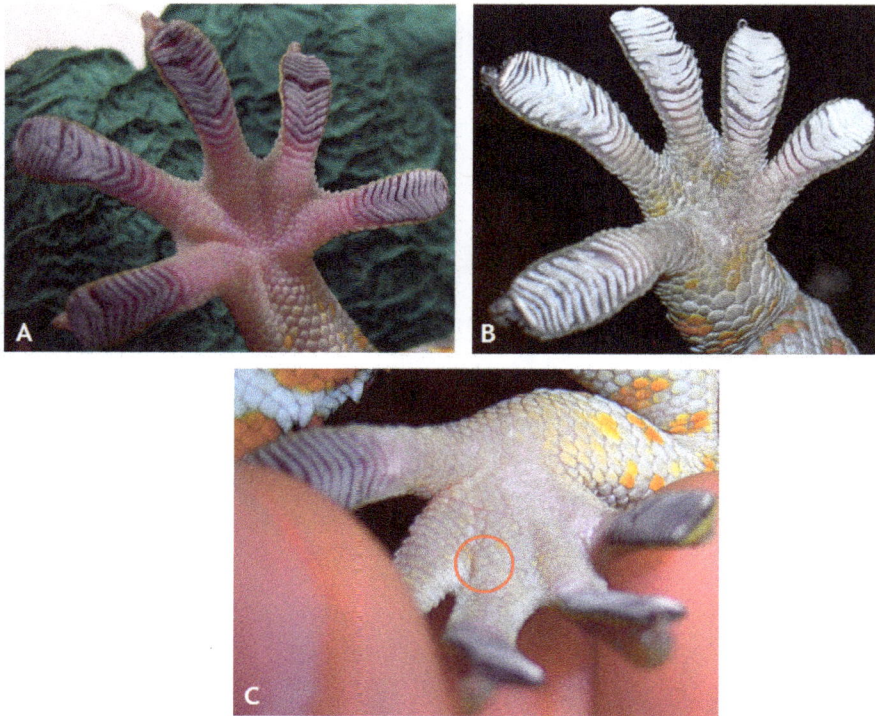

Figure 3.10 Damage imposed by the adhesive tests: (A) Diffused inflammation of gecko toes; (B) The gecko's healthy foot, for comparison; (C) Small, thin wound located on the gecko's skin between one toe and the next.

3.5. Conclusions

We have measured normal adhesive force-displacement curves for a live gecko. The gecko's maximum SF was determined to be λ_{PMMA} = 10.23 on PMMA surface, which shows higher roughness and index Sdr (25 times greater than that of glass), and λ_{Glass} = 9.13 on the glass surface. We observe a clear trend for adhesive ability after moulting: normal adhesive forces drastically decrease in each subsequent test as a consequence of the damage to the gecko's feet caused by previous experimental tests. Finally, we document the self-renewal of the gecko's adhesive system and abilities after moulting. The analysis reported here could also inform the design of bio-inspired smart adhesive materials.

Chapter 4

Optimal Angles for Maximal Adhesion in Living Tokay Geckos

Abstract

In this study, we report experimental measurements of the adhesion angles of living Tokay geckos (*Gekko gecko*) at two different characteristic sizes of the feet and toes. In particular, we have determined the adhesion angles between the opposing front and rear feet and between the first and fifth toe of each foot on various inverted surfaces (steel, aluminium, copper, Poly(methyl meth-acrylate) and glass). We explain the experimental results with the recently derived multiple peeling theory, and find an interesting agreement with previously reported observations on the architecture of the gecko adhesive system, even when considering the size scale of single seta, which suggests the validity of this approach at different hierarchical levels.

4.1. Introduction

Geckos and lizards usually climb in complex three-dimensional habitats, which necessitates the development of such sophisticated dry adhesive systems on their pads. During the last century, many secrets of the gecko's adhesion have been explained (Wagler, 1830; Simmermacher, 1884; Schmidt, 1904; Dellit, 1934; Ruibal, 1965; Hiller, 1968; Gennaro, 1969; Russell, 1975; Russell, 1986; Schleich, 1986; Irschick, 1996; Autumn, 2000; Gorb, 2001a; Autumn, 2002a; Autumn, 2002b; Arzt, 2003; Bergmann, 2005; Huber, 2005b; Gao, 2005; Hansen, 2005; Autumn, 2006b; Huber, 2007; Pugno, 2007a; Lepore, 2008; Pugno, 2008a; Pugno, 2008b; Lepore, 2010; Russell, 2002; Autumn, 2006b; Autumn, 2006c), although some crucial questions still remain unsolved (Autumn, 2002a; Gao, 2005; Autumn, 2006b; Gravish, 2008; Jusufi, 2008; Irschick, 2003; Tian, 2006). Such questions include: function, molecular mechanism, morphological characteristics of the nano-hierarchical structures, mechanism of frictional adhesion, tail function during climbing or aerial descent, and interactive effects of size and loading on kinematics. The millisecond-controllable attachment/detachment mechanism in geckos with negligible force has assumed a huge importance from a technological point of view, *e.g.* fabrication of dry adhesives, robotic systems, and artificial adhesive suits and gloves for astronauts (Autumn,

2006b; Pugno, 2007a; Autumn, 2006c; Lee, 2008a; Santos, 2007; Aksak, 2007; Murphy, 2007; Yao, 2008; Schubert, 2008; Shah, 2004). The uniqueness of the gecko's adhesive system, facilitated by repeatable strong feet contacts combined with temporary and reversible weak bonds, is based on intermolecular van der Waals forces (Autumn, 2002a; Autumn, 2006c; Irschick, 2003; Pugno, 2011; Zaaf, 2001; Russell, 2009; Haase, 1900). In order to maintain the necessary shear/frictional adhesive forces to avoid toe detachment (Autumn, 2000), the gecko's adhesive mechanism is based on the use of opposing feet and toes to form a V-shaped geometry. Attachment is achieved only proximally along the toe axis of the gecko, which pulls its feet inwards towards the center of mass (COM) and its toes inwards towards the foot to engage adhesion (Autumn, 2002a; Autumn, 2006b; Autumn, 2006c; Gravish, 2008; Tian, 2006; Haase, 1900; Pesika, 2007), as schematically reported in Fig.4.1A.

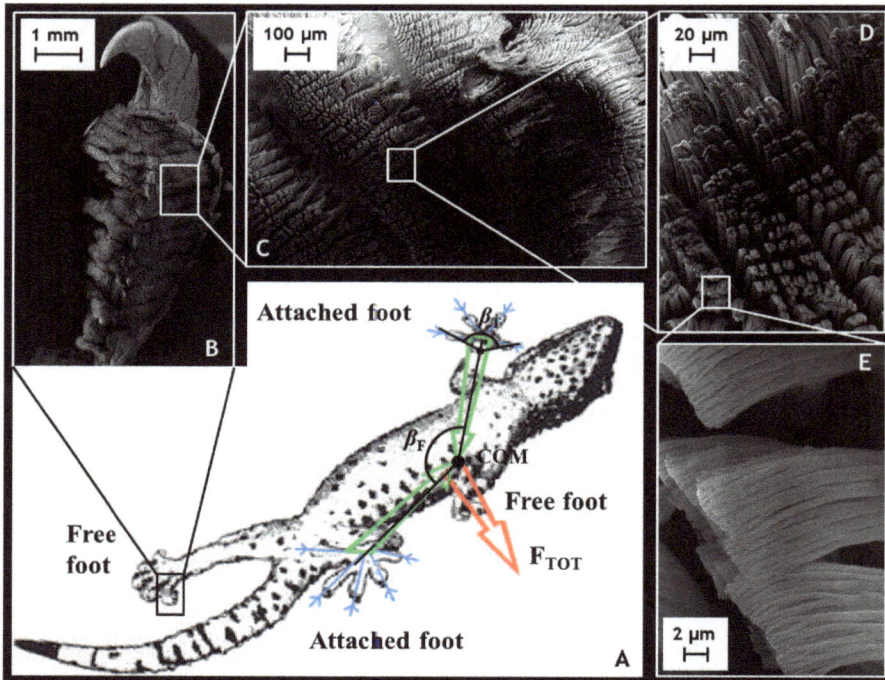

Figure 4.1 (A) A schematic 3D representation of the measured angle between the opposing front and rear feet (β_r) and between the first and fifth toe (β_r) of each foot on inverted surfaces (inset adapted from Y. Tian, N. Pesika, H. Zeng, K. Rosenberg, B. Zhao, P. M., K. Autumn, and J. Israelachvili, *Adhesion and friction in gecko toe attachment and detachment*, 19320-19325, PNAS, December 19, 2006, vol. 103, no. 51; Copyright (2006) National Academy of Sciences, U.S.A.). The Tokay gecko adhesive system observed using FESEM (Zeiss SUPRA 40) (B, C, D) and by SEM (Zeiss EVO 50) (E). The gecko's toe (B), FESEM micrograph of setae arrays (C), SEM micrograph of several setae (D) and nanoscale array of hundreds of spatula tips (E).

The key factor that governs the gecko's attachment/detachment is the adhesion angle α, which is formed between the terminal structure attached to the surface and the surface itself. Several studies have been performed to establish the value of the angle α from an experimental (Autumn, 2000; Autumn, 2006b; Gravish, 2008; Jusufi, 2008 Lepore, 2012c), computational (Autumn, 2002a; Gao, 2005; Gravish, 2008; Shah, 2004; Kim, 2007) or theoretical (Gao, 2005; Autumn, 2006b; Tian, 2006; Shah, 2004; Pesika, 2007; Pugno, 2011; Autumn, 2006c; Autumn, 2006a; Sitti, 2003; Spolenak, 2005; Yao, 2006; Pugno, 2008c; Varenberg, 2010; Buehler, 2006; Sen, 2010; Buehler, 2010) point of view and at different characteristic levels of the hierarchical adhesive system. From the literature, the angle α of Tokay geckos (*Gekko gecko*) is equal to ~25.5° for a single toe, ~24.6° (or ~30°) for isolated setae arrays, and ~30.0° (or ~31°) for a single seta (Autumn, 2006b) (or (Gravish, 2008)).

In this study, we experimentally evaluate the adhesion angles of living Tokay geckos at two different hierarchical characteristic sizes of the feet and toes. We measured the angles between the opposing front and rear feet and between the first and fifth toe of each foot on five different surfaces (steel, aluminium, copper, Poly(methyl meth-acrylate), *i.e.* PMMA, and glass). We compare the results with the recently published theory of multiple peeling (Pugno, 2011) and other previously published experimental results, yielding interesting findings. The results could be useful for the industrial fabrication of dry adhesives, robotic systems, artificial adhesive suits and gloves for astronauts, and the designi of bio-inspired smart adhesive nanomaterials.

4.2. Materials and Methods

We studied a single male adult Tokay gecko maintained in a terrarium at ~28 °C. The gecko was provided food (moths and crickets with a calcium supplement) and water *ad libitum*. The animal and the experimental procedures were authorized by the Ministerial Decree n° 73/2010-B.

The animal was placed in a natural position on the bottom of a box (50 cm x 50 cm x 50 cm). Each surface of the box was made of a different material (steel, aluminium, copper, PMMA, and glass) (Vetronova, Varese, Italy).

The box was slowly rotated until the gecko reached a downwards position under only its own weight (~88 g). At this time, we recorded the adhesion angle between the opposing front and rear feet (β_F) and between the first and fifth toe (β_T) of each foot on the inverted box surface. The legs are named as follows: front right (FR), front left (FL), rear right (RR), and rear left (RL). Experiments were performed at room (experimental box) temperature of ~21 °C (~25 °C) and humidity of ~50 % (~30 %). Figs.4.2 and 4.3 demonstrate how the adhesion angles between the opposing front and rear foot (Fig.4.2), on the

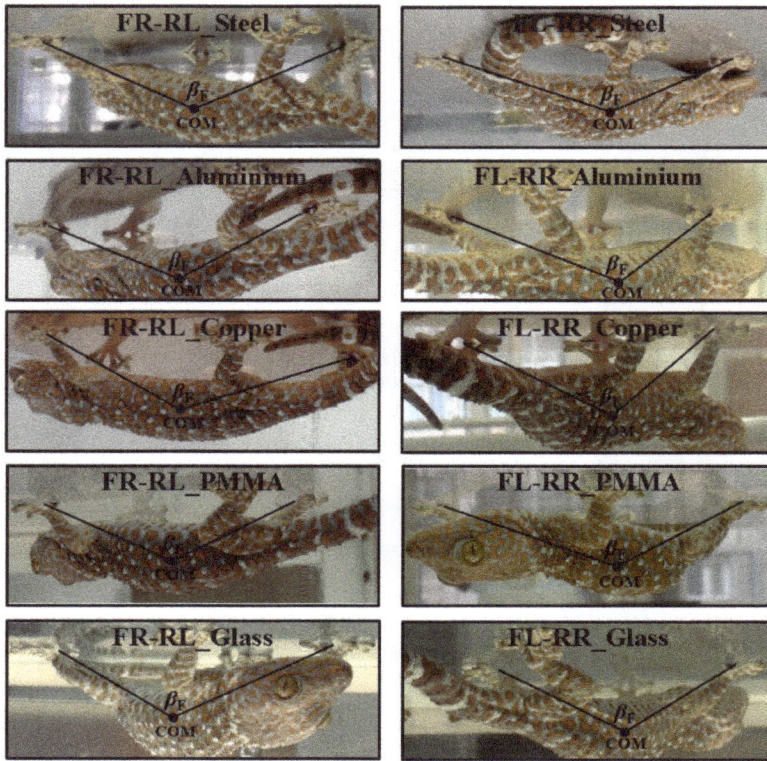

Figure 4.2 The measured angle β_F between the opposing front and rear feet on different surfaces.

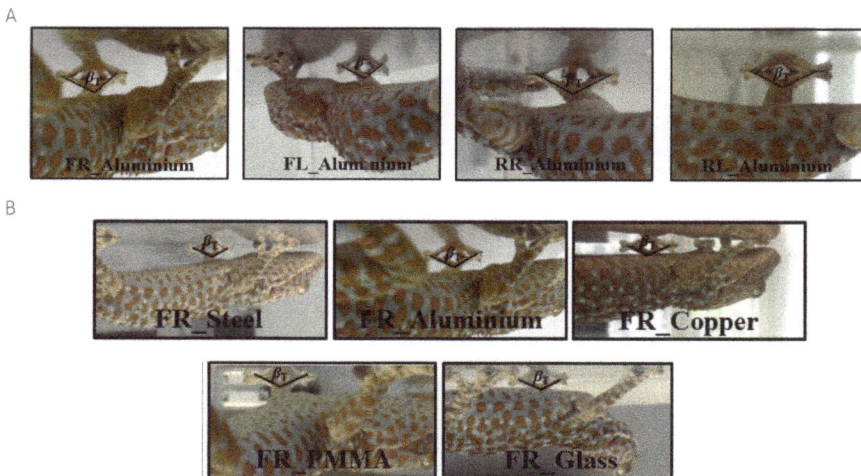

Figure 4.3 The measured angle β_T between the first and fifth toe: on the aluminium surface for all legs (A), or for the FR leg on different surfaces (steel, aluminium, copper, PMMA, and glass).

same substratum (Fig.4.3A), or of the same leg (Fig.4.3B) were measured. The angle β_T was determined between the first and fifth toe by taking the foot-forearm joint as the vertex of the resulting triangle. Similarly, the angle β_F was determined between the opposing front and rear feet by using the center of the gecko's mass (COM) as the vertex of the resulting triangle as defined in (Autumn, 2006e). The resulting angle α was computed as $\alpha=(180°-\beta)/2$.

4.3. Results

The experimental measurements of the adhesion angles are summarized in Table 4.1.

Table 4.1 Experimental values (and the number N of measurements) for the adhesion angles α_F (A) and α_t (B) on different surfaces.

A

α_F (°)	FR-RL	FL-RR	MEAN ± s.d.	λ_F
Steel	28 ± 4.7 (N=21)	29 ± 4.6 (N=39)	29 ± 0.6	0.013
Aluminium	31 ± 4.5 (N=33)	31 ± 4.2 (N=57)	31 ± 0.3	0.018
Copper	31 ± 8.0 (N=25)	33 ± 3.8 (N=37)	32 ± 0.8	0.023
PMMA	22 ± 4.5 (N=19)	26 ± 6.9 (N=33)	24 ± 2.8	0.007
Glass	28 ± 3.7 (N=22)	31 ± 3.8 (N=28)	30 ± 1.6	0.016
MEAN ± s.d.	28 ± 3.7	30 ± 2.4		

B

α_T (°)	FR	FL	RR	RL	MEAN ± s.d.	λ_T
Steel	28 ± 2.9 (N=24)	31 ± 4.3 (N=22)	31 ± 4.8 (N=44)	28 ± 4.4 (N=40)	30 ± 1.7	0.016
Aluminium	28 ± 3.9 (N=39)	30 ± 4.1 (N=50)	29 ± 4.3 (N=14)	28 ± 4.8 (N=28)	29 ± 0.8	0.013
Copper	24 ± 3.4 (N=30)	32 ± 6.1 (N=35)	28 ± 4.5 (N=43)	29 ± 4.7 (N=45)	28 ± 3.4	0.013
PMMA	21 ± 2.4 (N=24)	23 ± 2.7 (N=23)	21 ± 3.5 (N=27)	19 ± 2.0 (N=27)	21 ± 1.8	0.003
Glass	27 ± 3.7 (N=56)	32 ± 2.5 (N=37)	30 ± 3.3 (N=33)	27 ± 5.7 (N=18)	29 ± 2.5	0.015
MEAN ± s.d.	26 ± 3.0	29 ± 3.8	28 ± 4.3	26 ± 4.3		

It is clear that the FR leg value of α_T is lower than that of the FL leg for each surface and, similarly, the RL leg shows a lower value of α_T than the RR leg with the exception of the value on the copper surface. Moreover, the opposing FR and RL legs show the smallest values of α_T compared with the FL and RR legs. The determined values of α_F and α_T in Tokay geckos are in agreement with previous results reported by Autumn (Autumn, 2006e), which indicate the range of 25°-30° for α for toes, isolated setae arrays, or a single seta. This suggests a maximum in the gecko's attachment force when α reaches values around 30° (Santos, 2007; Shah, 2004).

4.4. Discussion

A good correlation was found between the experimental results and the theory of multiple peeling (Pugno, 2011), according to which the dimensionless detachment force of a V-shaped system is:

$$ f = \frac{F_C(\alpha)}{F_C(\pi/2)} = \frac{\sin\alpha \left(\cos\alpha - 1 + \sqrt{(1-\cos\alpha)^2 + 4\lambda} \right)}{-1 + \sqrt{1+4\lambda}} \tag{1} $$

where α is the adhesion angle and

$$ \lambda = \frac{\gamma}{tE}, \tag{2} $$

where γ is the surface energy, t is the tape thickness, and E is the Young modulus. Thus, when Eqn. 1 is applied to our data (with the mean values of α_F and α_T for each surface), we determine the dimensionless adhesion strength λ for the five surfaces at each hierarchical level (of foot and toe), graphically shown in Fig.4.4 and reported in the right column of Table 4.1. Note that λ_T is smaller than λ_F (except for the steel surface). Thus, as suggested by the multiple peeling theory, the smaller the parameter λ ($\lambda_T < \lambda_F$), the smaller the optimal adhesion angle ($\alpha_T < \alpha_F$), which corresponds to the peak value of the function f shown in Fig.4.4.

Following (Yao, 2006), we expect at each hierarchical level n the validity of the following equation:

$$ \frac{8}{(1-v_f^2)\pi} \frac{\gamma_n (E_n/\phi_{n-1})}{(\sigma_{th}\phi_{n-1})^2 R_n} = 1 \tag{3} $$

where $\gamma_n = W_n^{ad}$ is the work of adhesion, $E_n/\phi_{n-1} = E_f$ is the elastic modulus of a fiber, v_f is the Poisson's ratio of a fiber, $\sigma_{th}\phi_{n-1} = S_n = E_n\varepsilon$ is the effective adhesion strength, and $\phi_{n-1} = \prod_{i=1}^{n-1}\varphi_i$, where φ_i is the area fraction.

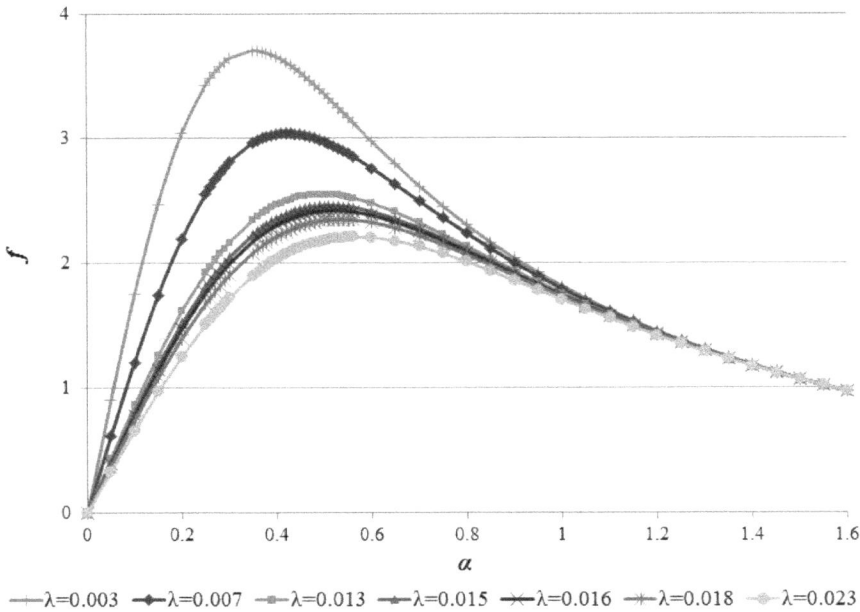

Figure 4.4 From the multiple peeling theory (Sitti, 2003), the dimensionless force *f versus* adhesion angle α using experimental mean values for α_F and α_T (fitting parameters λ reported in Table 4.1).

Thus, according to Eq. 3, Eq. 2 can be rewritten for each hierarchical level n as:

$$\lambda_n = \phi_{n-1},$$

(4)

which suggests an inverse dependence of the parameter λ on the number n of hierarchical levels, so the parameter λ decreases as n increases.

We define the Young modulus $E_T = E_F = 1$ GPa and the hierarchical level n and the thickness t of feet ($n_F = 4$ (four feet per gecko), $t_F = 10$ mm) and toes ($n_T = 5$ (five toes per each foot), $t_T = 4$ mm). Thus, using the work of adhesion γ of a previous study (Yao, 2006) (varying in the range of 10^3-10^6 J/m²), we find a theoretical range of λ ($10^{-4} \div 10^{-1}$) which confirms the previously computed values of the parameter λ ($10^{-2} \div 10^{-1}$) (Yao, 2006) and the experimental range of λ ($10^{-3} \div 10^{-2}$) determined here.

A further consideration concerns the critical angle α_c, which corresponds to the inclination of the force vector (F_{TOT} in Fig.4.1) just before the gecko's detachment as follows:

$$\alpha_c = arctg\left(\frac{F_n}{F_s}\right) \leq \alpha$$

(5)

where F_n and F_s are the normal and shear adhesive forces, respectively.

The experimental value of the critical angle α_c is ~11.3° for the *Hemidactylus garnotii* gecko (calculated with F_n = 0.006 N and F_s = 0.03 N (Autumn, 2006b)). For Tokay geckos, the critical angle α_c is ~9.5° at the level of the entire animal (calculated with F_n = 6.7 N (Fugno, 2011) and F_s= 40.2 N (Simmermacher, 1884)) or ~11.3° at the level of setae (calculated with F_n = 40 μN (Autumn, 2002b) and F_s = 200 μN (Autumn, 2000)). Note that the experimental values of the critical angle α_c confirm the previously reported range of 5.2°-11.3° for the whole insect (Autumn, 2006b), and are in agreement with Eq. 5, so they are coherently smaller than the optimal adhesion angle α, as experimentally determined here.

A final consideration regards the linear equation which fits experimental data of the perpendicular adhesive force F_n of the gecko's setae and the adhesion angle α (Autumn, 2000):

$$\alpha = \frac{0.22}{1N} \cdot F_n + 28.2.$$ (6)

Interestingly, when the normal adhesive force F_n= 6.7 N for the whole Tokay gecko (Pugno, 2011) was utilized in Eq. 6, the value α of 28.6° (or 28.9°) between the opposing first and fifth toe (or between the opposing front and rear feet) is obtained by roughly dividing F_n by 4 for the number of the gecko's feet (or by 2 for the number of couples of opposing front and rear feet). This result is in agreement with the experimental adhesion angles reported here.

4.5. Conclusions

In summary, the gecko's adhesion angle α has been estimated for a single toe (~25.5° (Autumn, 2006b)), for isolated setae arrays (~24.6° (Autumn, 2006b), ~30° (Gravish, 2008)) and for a single seta (~30.0° (Autumn, 2006b), ~31° (Autumn, 2000)). In this study, they are calculated for the angles between the opposing front and rear feet (a_{F_FR-RL} = 28°, a_{F_FL-RR} = 30°) and between the first and fifth toe of each foot (a_{T_FR} = 26°, a_{T_FL} = 29°, a_{T_RR} = 28°, a_{T_RL} = 26°) directly for the whole gecko (Pugno, 2011). Thus, angles in the range from ~26° to ~30° seem to be optimil to maximize surface adhesion for living Tokay geckos. The agreement between the theoretical calculations from the multiple peeling theory and the experimental results at the level of foot and toe evaluated here, in addition to those already reported in the literature about the gecko adhesive system (single toe, isolated setae arrays and single seta), support the validity of our approach at different hierarchical levels and provides an important contribution to the literature. The presented findings could be useful for the industrial fabrication of bioinspired dry adhesives tapes, robotics systems, artificial adhesive suits and gloves for astronauts, the design of bio-inspired smart adhesive nanomaterials, and even biomedical applications.

Chapter 5

Observations of Shear Adhesive Force and Friction of *Blatta Orientalis* on Different Surfaces

Abstract

The shear adhesive force of four non-climbing living cockroaches (*Blatta Orientalis* Linnaeus) was investigated using a centrifugal machine to determine of the shear safety factor on six surfaces (steel, aluminium, copper, two sand papers, and a common paper sheet). The adhesive system of *Blatta Orientalis* was characterized by means of a field emission scanning electron microscope, and the surface roughness was determined by an atomic force microscope. The presented findings highlight an interesting correlation between cockroach shear adhesion and the surface roughness with a threshold mechanism dictated by the competition between claw tip radius and roughness.

5.1. Introduction

The adhesive abilities of insects, spiders and reptiles have inspired scientists and researchers for a long time. These organisms present outstanding performance particularly for frictional and adhesive forces related to climbing abilities, especially considering their weight. In particular, it is well known how small insects can carry many times their own weight while walking quickly.

During the last few decades, a multitude of insects have been studied in order to understand and measure their adhesive abilities using microscopy instruments (Scanning Electron Microscope (SEM) and Atomic Force Microscope (AFM)), such as common beetles (Stork, 1980; Eisner, 2000; Voigt, 2008; Dai, 2002; Bullock, 2008; Eigenbrode, 2002); flies (Wigglesworth, 1987; Dixon, 1990; Lees, 1988; Gorb, 2001b; Walker, 1985; Voigt, 2005); ants (Federle, 2000; Federle, 2002; Brainerd, 1994; Federle, 2003); cockroaches (Arnold, 1974; Bell, 2007; Van Casteren, 2008; Clemente, 2008; Clemente, 2009 Lepore, 2013); spiders (Niederegger, 2006; Kesel, 2004; Kesel, 2003); and geckos (Autumn, 2002a; Pugno, 2011; Autumn, 2006b; Pugno, 2008a; Pugno, 2008b; Lepore, 2008; Huber, 2007; Irschick, 1996; Pugno, 2007a; Pugno, 2008c; Varenberg, 2010).

Biological adhesion can be obtained through different mechanisms (*e.g.* claws, clamp, sucker, glue, friction), although insect attachment pads have evolved in two main types; those which are hairy (thousands of flexible hairs, such as fly pulvilli and beetle pads) and those which are smooth (a highly deformable material, as for grasshoppers and cockroaches) (Peattie, 2009; Gorb, 2000). For example, geckos possess a dry adhesive surface organized in a hierarchical structure (Autumn, 2002a) like anoles, skinks, and spiders. Other animals possess secretion-aided fibrillae or pads, which are common in insects, especially hexapods (Beutel, 2001) like ants (Federle, 2002) and cockroaches (Arnold, 1974). The adhesive organs of these insects consist of smooth pads, and adhesion is mediated by a small volume of fluid secreted into the contact zone (Drechsler, 2006). In general, the adhesive structure and mechanism can be correlated with the micro-structured roughness of substrata (*e.g.* plant surfaces), which such animals interact with in nature (Gorb, 2002; Bitar, 2010) and thus have a strong influence on their adhesive abilities (Persson, 2007).

The normal and shear adhesive forces of several animals have been determined to assess their climbing abilities. As a matter of fact, these animals must endure not only perpendicular but also shear forces. For example, during the last few decades the adhesion of the Tokay gecko (*Gekko gecko*, ~100 g), which possesses the most widely studied biological adhesive system, was measured in terms of the normal force (Pugno, 2007a), the shear force (Autumn, 2006b), the adhesion time (Pugno, 2008a), and the influence of surface roughness on adhesive properties (Pugno, 2008b; Huber, 2007).

The shear adhesive force, and thus the shear safety factor (sSF) obtained by dividing the shear adhesive force by body weight, is an apparent friction coefficient that has already been determined for some living animals through various techniques (Van Casteren, 2008): with a centrifuge machine, the sSF was estimated to be ~70 and ~60 for male and female specimens of the Colorado potato beetle *Leptinotarsa decemlineata* (mass ~121 mg and ~168 mg, respectively) (Voigt, 2008), ~43 for syrphid flies (mass ~62 mg) (Gorb, 2001b), ~843 for the ant *Oecophylla smaragdina* (mass ~4 mg) (Federle , 2003), ~18 and ~14 for male and female specimens of the codling moth *Cydia pomonella* (mass ~19 mg and ~30 mg, respectively) (Bitar, 2010), and ~70 for the bug *Coreus marginatus* (mass ~80 mg) (Gorb, 2004); with the application of weights or force transducer, the sSF is equal to ~40 for the beetle *Pachnoda marginata* (mass ~1 g) (Dai, 2002), ~109 and ~3 for the leaf beetle *Gastrophysa viridula* and the stick insect *Carausius morosus* (mass ~10.4 mg and ~898 mg, respectively) (Bullock, 2005), ~28 for the blowfly *Calliphora vomitoria* (mass ~72 mg) (Walker, 1985), ~100 for geckos (mean mass ~10 g), ~60 for anoles (mean mass ~9 g), and ~18 for skinks (mean mass ~9 g) (Irschick, 1996), ~317 and ~81 for male and female specimens of the leaf beetle *Gastrophysa viridula* (mass ~10.8 mg and ~19.7 mg, respectively) (Bullock, 2009), and finally «1 for the green bushcricket *Tettigonia viridissima* (mass ~1 g, on polished silicon substratum) (Gorb, 2000).

We have focused on the shear adhesive force of living cockroaches (*Blatta Orientalis* Linnaeus), a species of the Blattodea order. There are thousands of species of cockroaches and only a few of these species live in human environments. The species of Blattodea are distinguished between climbing (*i.e. Blattella Germanica*) and non-climbing (*i.e. Blatta Orientalis*) depending on their ability to climb on smooth vertical (or upside down) surfaces, like Poly(methyl meth-acrylate) (PMMA), Poly(ethylene terephthalate) (PET), and sheet metals.

In this study, we measure of the sSF of non-climbing living cockroaches (*Blatta Orientalis* Linnaeus) on six surfaces with different roughnesses (two different sandpapers, common paper, steel, aluminium, and copper) using a centrifuge technique. Four cockroaches were studied to assess the sSF via three measurements per individual on each surface. This procedure guarantees consistent biomechanical data correlated to the surface roughness, which is quantified by AFM. The adhesive system of *Blatta Orientalis* was characterized by a field emission scanning electron microscope (FESEM) at the end of the experimental session.

5.2.Experimental Set-up

A self-built centrifuge device was used to measure the sSF of cockroaches. The centrifuge machine allowed us to avoid any prior treatment of cockroaches, which are left free of motion and assume a natural attachment position inside the experimental box. In addition, previous studies (Federle, 2000) indicate that the centrifuge device most likely yields higher values for adhesive forces than any other procedure of force measurement.

The sSF measurement depends only on the angular speed as the radius is constant (the position of the cockroach is far from the rotational axis). The experimental configuration is shown in Fig.5.1.

The box was built in order to facilitate an interchangeable floor to perform tests on different surfaces. The machine is composed of an electric motor (M_1 in Fig.5.1), which is used as a shaft, and another electric motor (M_2), which forces the system to rotate. M_1 and M_2 are connected through a belt transmission. M_2 is connected to a 220 V power source (50 Hz, AC) controlled with a frequency controller (VFD004L21E, Delta Electronics, named FC), which modulates the current frequency in the range of 1-400 Hz. Two aluminum bars are attached to the shaft and support the box B_1 of 25 cm x 25 cm x 25 cm on one side and the counterweight (CW) on the other side. The camera (C) is placed on the rotational axis (RA) of the system and records the cockroach's movement. Inside B_1 is another small box B_2 of 7 cm (w) x 4 cm (l) x 3 cm (h), where the animals were placed so that the uncertainty on the insect's radial position is reduced to ± 2.0 cm. The angular speed was measured with a standard bicycle computer (BCP-01, BBB company, named BC) by a magnetic sensor and LCD screen, which is fixed

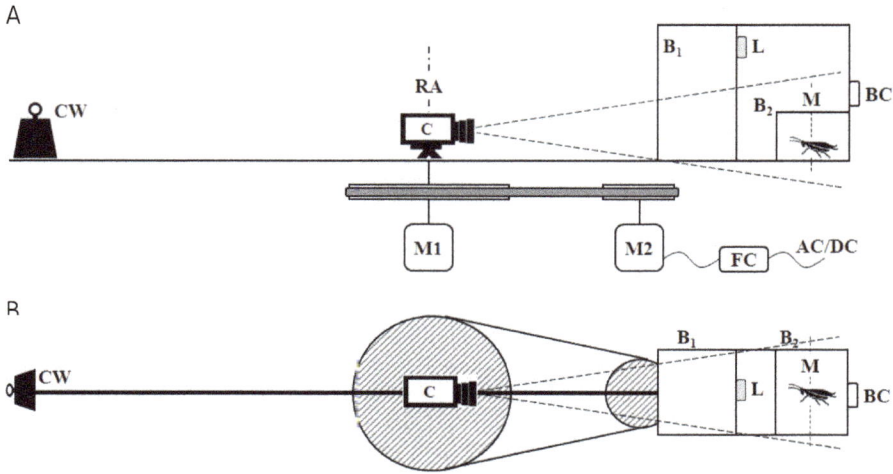

Figure 5.1 Side (A) and top (B) view of the centrifugal machine used to measure the insects sSF
(M1: passive rotating linchpin; M2: electric motor connected to M1 with a transmission belt; FC:
frequency controller to set the M2 rotational speed; RA: rotational axis; C: camera; B1: external box;
B2: internal small box where specimens were placed in; M: middle of the internal box; L: lamp; BC:
bicycle computer; CW: counterweight).

to the radially external wall of B_1. In the movie (in Fig.5.1, the blue lines identify the video shot), both the cockroach and the speed measurement were recorded so that the correct speed corresponding to the cockroach's detachment is determined. The BC was calibrated with the reference distance (51 cm) between the rotational axis and the middle (M) of B_2. The angular speed was calculated from the linear speed read on the BC LCD screen, which gives the speed value in the range of 0.0-199.9 km/h with an accuracy of ± 1 % over the read value.

To minimize the cockroach experience of the rotation, the box was insulated from the environment using a dark paper to obscure the box and adding an artificial light (L) inside B_1.

Experiments were conducted on four adult cockroaches (B_1, B_2, B_3, and B_4) of the species *Blatta Orientalis*. They were kept alone and fed chicken feed *ad libitum*. The insects were maintained at a room temperature of ~25 °C and humidity of ~50 %, which were also the experimental conditions.

The sSF measurements were conducted as follows. Four cockroaches were studied to assess the sSF using three measurements per individual on each surface. During the biomechanical experiments, a slow speed-up was provided to avoid high acceleration, which can facilitate an early detachment, and to satisfy the requirement of a constant angular speed in order to correctly evaluate the sSF. Every time the cockroach was placed on the bottom of the box, a two-minute time period was allotted so that it could familiarize with the room. During an acceptable test, the animal can still run on the bottom of B_2 at low speeds, whereas it walks more slowly as the centrifuge speeds up; finally, it stands still and contacts the substratum with all legs, assuming a 'freezing' position, which is advantageous to its attachment (Federle, 2000). By standing motionless with all legs spread out, the cockroach assumes a position that maximizes its adhesive ability, and detachment is not caused by natural movement but rather by the shear force acting on the animal. During an unacceptable test, the animals tended to go in a corner or against a wall, representing an aborting condition of the test. No animal was tested over more than two surfaces per day. The body weight of the four insects (equal to 405.9 ± 22.9 mg) was measured by a balance (EB200, Orma) with a precision of ± 0.1 mg.

5.3. Video Output

Fig.5.2 shows an example of two subsequent frames extracted from a single test video. The cockroach's detachment and tangential speed on the LCD screen are clearly visible. The cockroach stands on the surface until it detaches and goes immediately against the back wall due to the centrifuge force.

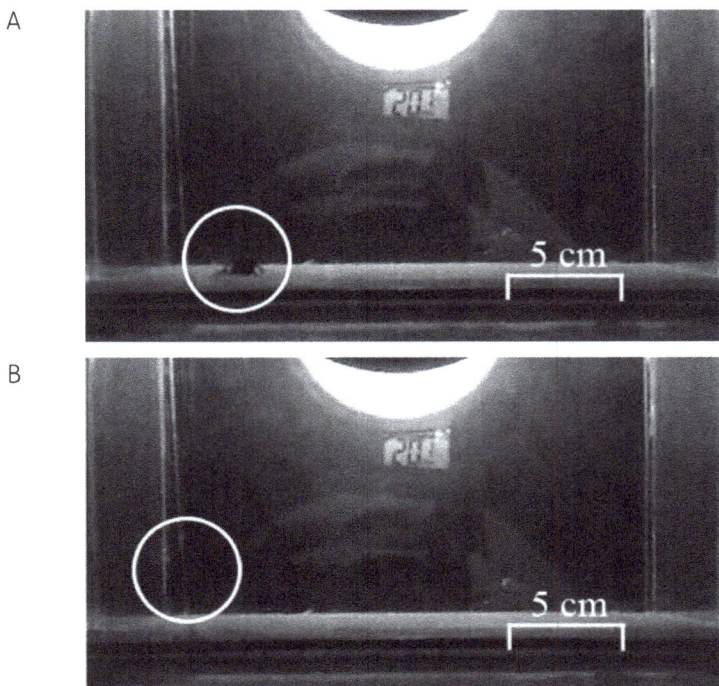

Figure 5.2 Two subsequent frames from a video: before detachment, the insect stands still on the surface (A) and, one frame later, it is in the box corner (B). These frames are extracted from a preliminary video without the use of the small box (B$_i$).

5.4. AFM Characterization of Surfaces

The characterization of surfaces (sheet of common office paper (80 g/m², named Cp), steel, aluminum, and copper) was performed in 'contact mode' with an AFM (Solver Pro M) with NSG01 tips (from NT-MDT, Moscow, Russia (Fig.5.3)). The parameters tuned during the analysis were the measurement speed (14.2 µm/s) and the measured area (100 µm x 100 µm for 3 tests on metals and 50 µm x 100 µm for 6 tests on Cp) with a final resolution of 512 points/profile. All parameters were referred to a 100 µm cut-off. The cut-off length defines the length from which the roughness parameters were calculated and therefore strongly influences the roughness values. The roughness parameters were determined with NOVA software from NTMDT (Moscow, Russia). No roughness data was obtained for the two types of sandpaper (Sp50 and Sp150) because their roughness is beyond the working ranges of the AFM; the mean nominal diameter of surface asperities was used to compare them with the AFM measured surfaces. See previous studies (Pugno, 2011; Pugno, 2008a; Pugno, 2008b; Lepore, 2008) for a detailed explanation of the classical roughness parameters (*Sa, Sq, Sp, Sv, Sz, Ssk*). *Ska* is the kustosis parameter and indicates the distribution of surface heights: when it is close to 0, the distribution of surface heights is like a Gaussian distribution; when it is higher than 0, the height distribution is sharper than a Gaussian distribution (so peak heights are close to the mean height); when is it lower than 0, the height distribution is more spread.

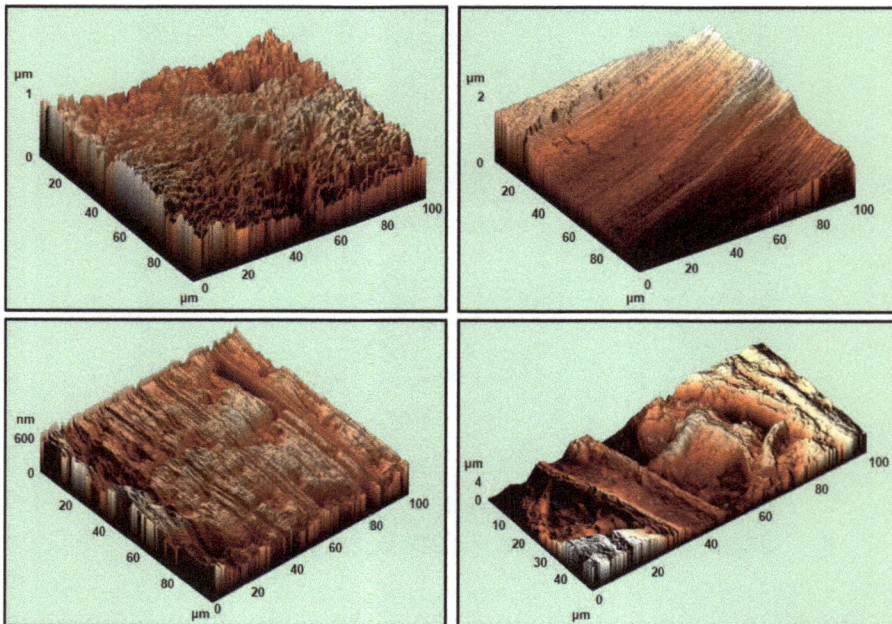

Figure 5.3 AFM characterization of the (A) steel, (B) aluminium, (C) copper, and (D) Cp surfaces.

5.5. FESEM Characterization of *Blatta Orientalis*

We observed the adhesive system of *Blatta Orientalis* by means of a FESEM (FEI-Inspect™ F50) equipped with a field emission tungsten cathode at 1 kV. Samples were amputated from naturally dead adult cockroaches and maintained in 70 % ethanol solution, 12-h dehydrated, fixed to aluminium stubs by double-sided adhesive carbon conductive tape (Nisshin EM Co. Ltd.), and scanned without metallization.

Fig.5.4 confirms the adhesive system description recently reported (Bell, 2007): a sub-obsolete nonfunctional arolium (not adapted for climbing a smooth vertical surface) with two claws for each of the six legs of *Blatta Orientalis*. The claw tip diameter was determined using the ImageJ 1.41o software to be 12.3 ± 4.73 μm.

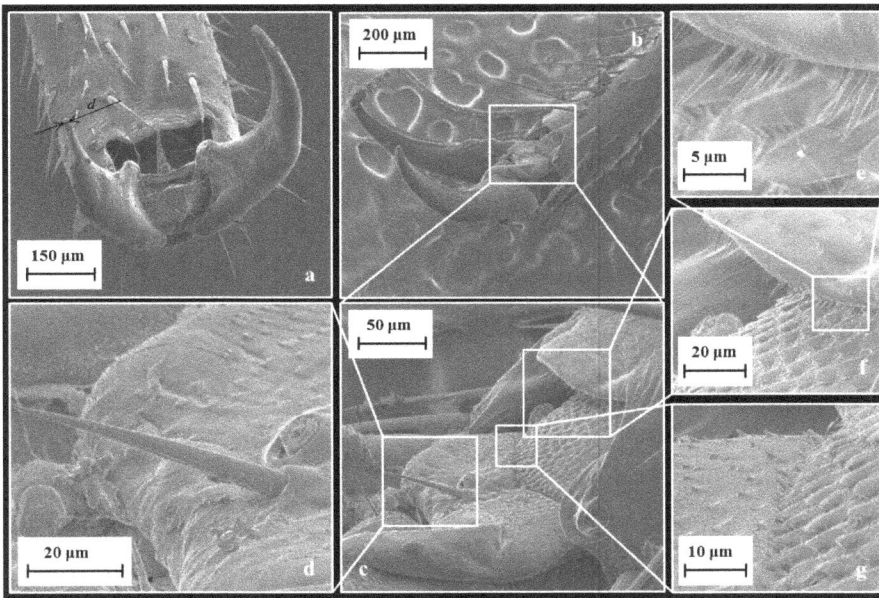

Figure 5.4 The adhesive structures of the legs of *Blatta Orientalis*. (a) Frontal and (b) lateral view of a leg and some detailed micrographs (c, d, e, f, g) (*d* is the claw tip diameter).

5.6. sSF Evaluation

The sSF, which is defined as the ratio between the shear detachment force ($F_{detachment}$) and the mass (m) multiplied by the gravity acceleration (g),is dimensionless and represents the apparent friction coefficient:

$$sSF = \frac{F_{detachment}}{m \cdot g} \tag{1}$$

We focus on the shear adhesive force and thus only consider the radial force (F_{radial}) acting on the insect. Thus, in our case $F_{detachment} = F_{radial}$. A constant angular speed (ω) is considered, so the radial force is proportional to the insect distance from the axis (the radius, $R = 51$ cm), the square of the angular speed, and the insect mass:

$$F_{radial} = m \cdot \omega^2 \cdot R \tag{2}$$

Thus, the sSF is easily calculated from:

$$sSF = \frac{\omega^2 \cdot R}{g} \tag{3}$$

Note that the sSF does not depend on the mass of the insect. Knowing the radius (constant) and neglecting the drag force as the insects are in a closed box, the sSF is obtained simply from the value of the angular speed measured on the BC.

5.7. Experimental Results

Fig.5.5 shows that there is no significant difference among the sSF of different insects. Thus, we simply average the results of the four tests on each surface. Table 5.1 reports the sSF and the F_{radial} for each surface (mean ± st.dev.) and shows a clear separation between rough (Sp50, Sp150, Cp) and smooth (steel, aluminium, copper) surfaces.

5.8. Discussion

In general, claw-mediated adhesive insects can attach to a horizontal or vertical surface just by interlocking, so adhesion increases with surface roughness (Bullock, 2008; Van Casteren, 2008) in agreement with our observations. In particular, claw-mediated adhesion occurs when the surface asperity size is comparable to or larger than the claw tip diameter (Voigt, 2008; Dai, 2002; Bitar, 2010), which is estimated to be 12.3 μm for *Blatta Orientalis*. Table 5.1 summarizes the calculated and estimated roughness parameters.

The unmeasured asperity diameter (*Ad*) for Cp, steel, aluminium, and copper (marked with [*] in Table 5.1) is estimated by multiplying the parameter *Sq* by the value of 3.6, the mean value *Sq/Ad* for sandpapers (Sp) from previously

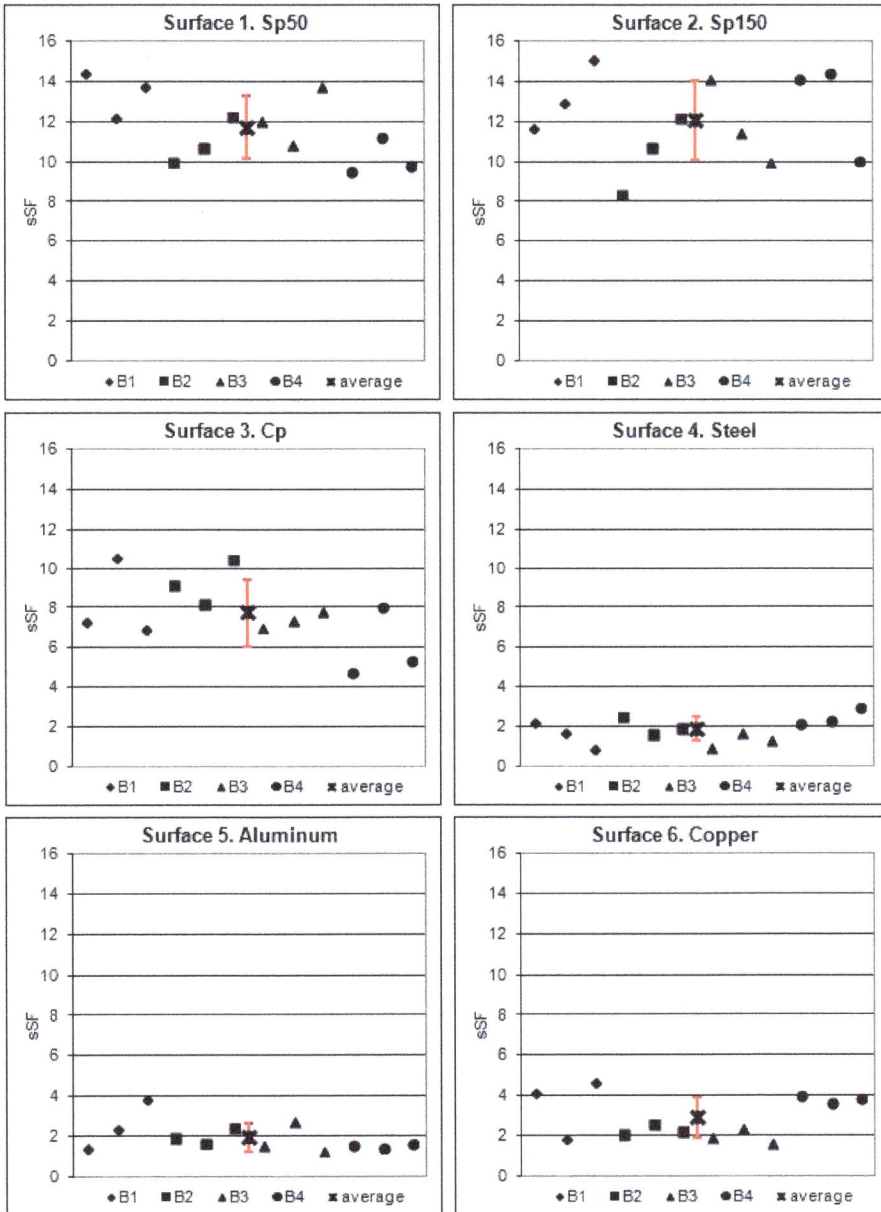

Figure 5.5 The sSF for each individual are grouped by surfaces.

Table 5.1 The roughness parameters, sSF and F_{radial} of the characterized insect/surface systems. The values [*] are computed by multiplying the parameter Sq by the value 3.6, which was previously calculated as Sq/Ad for sandpapers (Sp) with known Ad on which the roughness parameter Sq has been observed.

	Sp50	Sp150	Cp	Steel	Aluminium	Copper
Ad (µm)	336	100	4.5[*]	0.7[*]	0.6[*]	0.8[*]
S_a (µm)	-	-	1.044 ± 0.228	0.145 ± 0.041	0.141 ± 0.026	0.178 ± 0.125
S_q (µm)	-	-	1.248 ± 0.255	0.190 ± 0.053	0.173 ± 0.026	0.215 ± 0.145
S_p (µm)			2.727 ± 0.433	0.801 ± 0.176	0.626 ± 0.045	0.496 ± 0.258
S_v (µm)			3.132 ± 0.112	0.885 ± 0.353	0.434 ± 0.105	0.670 ± 0.237
S_z (µm)	-	-	2.927 ± 0.233	0.838 ± 0.190	0.521 ± 0.051	0.584 ± 0.228
S_{sk}	-	-	-0.31 ± 0.143	-0.78 ± 0.472	0.41 ± 0.331	-0.48 ± 0.590
S_{ka}	-	-	-0.66 ± 0.327	1.31 ± 0.485	-0.08 ± 0.820	-0.04 ± 1.139
sSF	11.7 ± 1.6	12.1 ± 2.0	7.7 ± 1.7	1.9 ± 0.6	2.0 ± 0.7	2.9 ± 1.0
F_{radial} (mN)	46.8 ± 8.5	48.1 ± 9.0	30.9 ± 7.2	7.4 ± 2.1	7.9 ± 3.2	11.6 ± 4.3

published papers (the value of Ad is known for sandpapers on which the roughness parameter Sq was calculated). It is reported here as Sp_{Ad}-Sq, thus $Sp_{30µm}$-6.66 µm, $Sp_{16µm}$-3.75 µm, $Sp_{12µm}$-3.25 µm, $Sp_{0.5µm}$-0.13 µm and $Sp_{1µm}$-0.4 µm (Bullock, 2010), $Sp_{12µm}$-3.06 µm, $Sp_{9µm}$-2.45 µm, $Sp_{3µm}$-1.16 µm, $Sp_{1µm}$-0.24 µm and $Sp_{0.3µm}$-0.09 µm (Huber, 2007), $Sp_{12µm}$-3.0603 µm, $Sp_{9µm}$-2.4537 µm, $Sp_{3µm}$-1.1567 µm, $Sp_{1µm}$-0.2384 µm and $Sp_{0.3µm}$-0.09 µm (Voigt, 2008). The results confirm our assumptions: the claws of *Blatta Orientalis* are not able to grip surfaces with Ad smaller than ~12 µm. Also, when compared with the shear adhesive forces on Sp50 and Sp150, a decrement of the shear adhesive forces of about 35 % on the Cp surface and of more than 80 % on metals is recorded.

Referring to the roughness analysis of steel, aluminium, copper, and Cp (Fig.5.3 and Table 5.1), it is clear that the Cp surface is characterized by the parameters Sa, Sq, Sp, Sv, and Sz of one order of magnitude higher than those of metal surfaces with a spread distribution of peak heights (Ska < 0), whose number is exceeded by the number of valleys (Ssk < 0), which are deep, wide and so most likely complementary to the geometry of the claw tip.

Referring to the metal surfaces, a noteworthy difference clearly emerges between copper and aluminium when compared with steel (on which we record the lowest sSF). The steel surface is denoted by a higher number of valleys than peaks ($Ssk < 0$), whose heights are very close to the mean value ($Ska > 0$) and which are usually at a distance lower than 1 μm. Thus, the lowest performance of *Blatta Orientalis* on the steel surface could be explained by the objective impossibility of the cockroach to interlock its claws inside the peak to peak distance.

The aluminium and copper surfaces are comparable for all of the roughness parameters with the exception of *Ssk*, which highlights that the cockroach *Blatta Orientalis* endures higher sSF on surfaces with a lower number of peaks than valleys, which most likely facilitate interlocking for its claws.

5.9. Conclusions

We measured the sSF of four non-climbing living cockroaches (*Blatta Orientalis* Linnaeus) using a centrifuge technique on six surfaces (two different sandpapers, common paper, steel, aluminium, and copper). The cockroach's maximum sSF, or apparent friction coefficient, is determined to be 12.1 on Sp150 ($Ad \approx 100$ μm, $F_{radial} = 48$ mN), and the minimum sSF is determined to be 1.9 on steel ($Ad \approx 0.7$ μm, $F_{radial} = 7.4$ mN). An interesting threshold mechanism is demonstrated between the cockroach's shear adhesive force and the surface roughness. It is clear that the best adhesion is obtained for roughness larger than the claw tip radius; surfaces with a higher number of valleys than peaks ($Ssk < 0$) and a spread distribution of peak heights ($Ska < 0$) also allow better adhesion.

Anti-Adhesive Materials

Chapter 6

Plasma and Thermoforming Treatments to Tune The Bio-Inspired Wettability of Polystyrene

Abstract

This study shows the effects on wettability of plasma and thermoforming treatments on fourteen different polystyrene (PS) surfaces as compared, with a lotus leaf. Quantitative roughness analyses of PS surfaces and lotus leaf by three-dimensional optical profilometer and scanning electron microscope have been performed. Water droplet sliding was characterized by measuring the contact angle, sliding angle, sliding volume, and sliding speed. A correlation between technological treatment, surface roughness parameters and wetting measurements emerges, which suggests that the plasma/thermoforming treatment enhances the hydrophilic/hydrophobic behaviour of PS surfaces. Our study concludes with determination of the static and resistant forces of droplet sliding on the studied surfaces.

6.1. Introduction

Water-repellent (or superhydrophobic) and dirt-free (or self-cleaning) natural surfaces might have been observed for the first time more than 2000 years ago. However, only in the 20th century have scientists studied these two related phenomena on natural leaves (Barthlott, 1981; Neinhuis, 1997; Barthlott, 1997; Herminghaus, 2000; Wagner, 2003; Nosonovsky, 2005; Lu-quan, 2007; Otten, 2004; Zhiqing, 2007; Brushan, 2008), *e.g.* the famous lotus *Nelumbo nucifera*, on which "raindrops take a clear, spherical shape without spreading, which probably has to be ascribed to some kind of evaporated essence", as Goethe described in 1817 (Solga, 2007).

In contrast to Goethe's conjecture, the so called lotus-effect is governed more than by chemistry (Young's law (Young, 1805)) by topology (Wenzel's law (Wenzel, 1936), Cassie-Baxter's law (Cassie, 1944)) and hierarchical architectures (Pugno, 2007a; Pugno, 2007b) (similar to what we observe for the strength and toughness of materials (Bosia, 2010, Bignardi, 2010; Pugno, 2006; Pugno, 2008d, Pugno, 2008e)). The contribution of surface roughness on superhydrophobic/

self-cleaning behaviour has been extensively shown in the literature (Burton, 2005; Bhushan, 2007a; Bhushan, 2007b; Jung, 2007a; Jung, 2007b; Nosonovsky, 2006; Nosonovsky, 2007d; Shibuichi, 1996; Hozumi, 1998; Miwa, 2000; Oner, 2000; Lau, 2003; Quéré, 2002). However, in some applications, materials should be hydrophilic more than hydrophobic, *e.g.* to maximize wettability.

In this study, we examine the effects of plasma and thermoforming treatments on different polystyrene (PS) surfaces. We have considered seven PS surfaces before (A_p) and after (B_p) the plasma treatment and fourteen PS surfaces before (A_t) and after (B_t) thermoforming treatment. All surfaces were analysed with a three-dimensional optical profilometer and a field emission scanning electron microscope. The hydrophilic behaviour given by plasma treatment is quantified by depositing distilled water droplets on PS horizontal surfaces with controlled and random volumes, showing a correlation between surface roughness parameters and contact angle (CA) measurements, in agreement with Wenzel theory. The effects of thermoforming treatment are quantified by measuring the droplet contact angle, sliding angle, sliding volume and speed. Finally, the static and resistant forces of a droplet sliding on each the surfaces is determined.

6.2. Materials and Methods

6.2.1. Plasma Treatment

A commonly applied method to increase wettability and chemical reactivity of polymeric materials by raising surface energy is plasma discharge treatment (also known as corona treatment). This treatment, invented by the Danish engineer Verner Eisby in the 1950s, is particularly suitable for continuous production processes like the one required for synthesis of the extruded PS sheets constituting the subject of the present study. This method is also safe, economical, and capable of high line-speed throughput.

Corona treatment is based on a high-frequency and high-voltage electrical discharge. The discharge is generated between an electrode and counter electrode. The corona discharge has such a powerful impact on the substance surface that the molecular structure changes in a way that improves surface wettability. A high voltage discharge inside an air gap causes air ionization. When a plastic material is placed in the discharge path, the electrons generated in the discharge impact the surface with energies two or three times larger than that necessary to break the molecular bonds. This creates reactive free radicals that, in presence of oxygen, can react rapidly to form various functional groups on the substrate surface. An evolution of this system, which is particularly efficient for higher activation potentials, is the plasma jet system, in which a high-voltage discharge (5-15 kV, 10-100 kHz) generates a pulsed electric arc. A process gas, usually oil-free compressed air flowing past the discharge site, is excited and converted to the plasma state. This plasma then passes through a jet head to arrive on the substrate surface. The jet head is at earth potential and in this way holds back a lot of the potential carried by the plasma stream. Corona surface and plasma jet treatment modifies only the surface characteristics without affecting material bulk properties (Noeske, 2004; Valeri, 1999; Bellucci, 2007).

Corona discharge treatment is commonly applied in the cooling appliance industry: refrigerator insulation systems are typically constituted by polyurethane foam and reticulated *in situ* within cavity designed by purpose. To ensure mechanical and thermal stability of the final assembly, and as a consequence of the modified surface topology due to plasma treatment, adhesion of polyurethane foam over surrounding surfaces, *i.e.* PS liner surface and external case, must be maximized. For the purposes of the present study, PS extruded slabs have been treated with the industrial "Ferrarini and Benelli" corona discharge system integrated within the refrigerator production line at Indesit Company. The main characteristics of the equipment are: nominal power (7.3 kVA), corona discharge power (6.5 kW), corona discharge device working frequency (30 kHz), achievable surface energy after treatment ($(4.2-5.6) \cdot 10^{-2}$ N/m), material temperature in treatment area (80 °C), performance test method (ASTM Standard Test Method D2578-84, "Wetting Tension of Polyethylene and Polypropylene Film").

6.2.2. Thermoforming Treatment

Thermoforming is a technology almost universally applied for refrigerator cabinet liner and door internal surface manufacturing; such a technique allows high throughput production and a good net shape surface finishing. Main phases of the process are: pre-heating (100°C), peak temperature (180°C), final temperature (70 °C).

After thermoforming, thickness reduction can exceed 90 % in some areas: careful control is required to verify that the sheet is kept robust (*e.g.* no breakage of aesthetic or functional layer), tuning the process and the material characteristics.

6.2.3. Surface Characterization

The characterization of PS surfaces was performed with a three-dimensional optical profilometer (Talysurf CLI 1000) equipped with the CLA Confocal Gauge

300HE or a mechanical cantilever with 300 μm range and 10 nm vertical resolution or with 546 μm range and 10 nm vertical resolution (Taylor Hobson, Leicester, UK). The parameters tuned during analysis are the measurement speed equal of 200 μm/s, the return speed of 1 mm/s or 500 μm/s, the sampling rate of 150 Hz or 40 Hz, the measured area of 500 μm x 500 μm and the resolution in the "*xy*" plane equal to 2.5 μm, leading to a final resolution of 201 points/ profile. All parameters were referred to a 250 μm cut-off. See paragraph 1.2 for a detailed explanation of the extracted classical roughness parameters (*Sa, Sq, Sp, Sv, Sz, Ssk, Sdr*) (Lepore, 2008; Pugno, 2008a; Pugno, 2008b).

The PS surfaces and lotus leaves were also observed by means of a field emission scanning electron microscope (FESEM, Zeiss SUPRA 40 for A_p, B_p, and A_t samples and lotus leaves, and FEI-Inspect™ F50 for B_t samples) equipped with a field emission tungsten cathode. Samples of ~1 cm² were obtained, fixed to aluminium stubs by double-sided adhesive carbon conductive tape (Nisshin EM Co. Ltd.), ethanol-cleaned (except for lotus leaf used as-is) and air-dried. Samples A_p, B_p, and lotus leaves or A_t and B_t were chrome or gold-coated, approximately 8 or 3.6 nm.

6.2.4. CA Measurement

The wettability of PS surfaces and lotus leaf was determined by measuring the static CA of distilled water droplets over samples fixed to a horizontal plane by double-sided adhesive tape and cleaned with ethanol before droplet deposition (in order to reduce the negative influence of sample cleanliness on contact angle measurements (Adamson, 1990; Israelachvili, 1992; Brushan, 1999; Brushan, 2002)). We consider a series of ten random-volume droplets gently deposited on the substrate with a standard single-use syringe and nine controlled-volume droplets (0.5, 0.7, 0.9, 1.1, 1.3, 1.5, 1.7, 1.9, 2.0 μl) deposited with a digital micropipette (Gilson, Ultra-range U2-Model, 0.2-2.0 μl). The contact angle was recorded with an OLYMPUS MJU 1010 digital photocamera, measured, and statistically analysed with the software ImageJ 1.41o.

6.2.5. Sliding Measurements

Two conceptually distinct procedures were used to evaluate the sliding angles on B_t samples and lotus leaves: (1) fixing the volume of the droplet (~16 μl) and measuring the angle of the sample stage at sliding or (2) fixing the angle of the specimen stage (90°) and measuring the sliding volume.

6.3. Results

6.3.1. Surface Characterization

Table 6.1 summarizes the extracted roughness parameters from the profilometer, and Fig.6.1 shows the related FESEM images (surface morphologies at the same magnification) of all PS materials. Figs.6.2-6.4 show the untreated PS surfaces, and Fig.6.5 shows the typical topography of plasma-treated samples. Fig.6.6 shows the effects of thermoforming treatment on representative samples (1 and 4). Fig.6.7 displays the profiles extracted at 50 mm from the edge of the square measured area. Finally, the SEM morphology of the adaxial leaf surface of the water-repellent and self-cleaning lotus is reported in Fig.6.8.

6.3.2. CA Measurement

In Table 6.2, the mean values and standard deviation of nineteen CA measurements for each PS surface are reported.

6.3.3. Sliding Measurements

The results of the first procedure for the determination of sliding angle show that PS surfaces have a sliding angle greater than 90° (no sliding). An exception is shown by sample $4B_t$, which shows a sliding angle of 48 ± 15.7 ° (Fig.6.9). The sliding volume V_s and the sliding speed v_s for B_t surfaces were determined by means of the second procedure. The values of V_s and v_s were calculated from five measurements per each sample, see Fig.6.10.

Table 6.1 Measured roughness parameters of PS surfaces. Note that samples $7A_p$ and $7B_p$ are used only to evaluate the effects of plasma treatment, whereas $7A_t$ and $14A_t$ are new starting samples for the determination of the effects of thermoforming treatment.

	S_a (μm)	S_q (μm)	S_p (μm)	S_v (μm)	S_z (μm)	S_{sk}	S_{dr} (%)
$1A_p=1A_t$	0.671±0.014	0.859±0.016	4.267±0.309	5.340±1.282	8.240±0.689	-0.274±0.012	5.583±0.104
$2A_p=2A_t$	0.753±0.049	0.970±0.060	4.697±0.926	6.027±0.195	8.967±0.320	-0.136±0.043	5.277±0.293
$3A_p=3A_t$	0.205±0.006	0.266±0.004	1.907±1.025	1.803±0.550	2.790±0.649	0.217±0.181	0.273±0.036
$4A_p=4A_t$	0.086±0.009	0.126±0.016	2.160±0.691	1.863±0.861	2.833±0.578	0.821±0.099	0.108±0.024
$5A_p=5A_t$	1.197±0.120	2.143±0.159	11.500±1.179	16.500±1.969	24.867±0.723	-2.523±0.483	11.003±1.731
$6A_p=6A_t$	0.120±0.018	0.156±0.025	1.201±0.309	0.896±0.196	1.530±0.399	0.138±0.124	0.060±0.022
$7A_t$	0.744±0.084	0.946±0.115	3.553±1.703	4.613±0.664	6.01±0.840	-0.444±0.178	0.486±0.093
$1B_p=8A_t$	1.730±0.095	2.203±0.125	12.963±5.797	13.927±6.907	21.300±3.863	-0.074±0.132	20.800±1.345
$2B_p=9A_t$	1.330±0.056	1.693±0.078	6.960±0.259	9.353±1.021	14.167±0.808	-0.144±0.159	14.500±0.794
$3B_p=10A_t$	0.921±0.009	1.187±0.011	4.403±0.195	6.657±0.447	10.080±0.470	-0.311±0.099	7.090±0.157
$4B_p=11A_t$	1.427±0.068	1.857±0.075	8.747±0.473	9.780±0.121	16.400±0.600	-0.383±0.187	12.967±0.929
$5B_p=12A_t$	0.939±0.030	1.213±0.035	5.627±1.137	6.733±1.259	10.473±1.040	-0.289±0.205	6.293±0.780
$6B_p=13A_t$	1.273±0.136	1.653±0.182	6.657±0.631	9.553±0.624	14.367±0.924	-0.396±0.110	11.663±1.807
$14A_t$	0.745±0.132	0.953±0.166	4.555±0.912	4.425±1.365	6.365±0.502	-0.171±0.064	0.617±0.057
$7A_p$	0.313±0.016	0.403±0.023	2.647±0.894	2.383±0.365	3.913±0.682	0.076±0.118	0.629±0.120
$7B_p$	1.427±0.176	1.867±0.244	20.757±17.423	11.333±0.751	24.767±11.829	-0.306±0.449	15.533±3.060

continued **Table 6.1** Measured roughness parameters of PS surfaces. Note that samples $7A_t$ and $7B_t$ are used only to evaluate the effects of plasma treatment, whereas $7A_t$ and $14A_t$ are new starting samples for the determination of the effects of thermoforming treatment.

	S_a (μm)	S_q (μm)	S_p (μm)	S_v (μm)	S_z (μm)	S_{sk}	S_{dr} (%)
$1B_t$	0.841 ± 0.201	1.059 ± 0.240	3.443 ± 1.364	3.970 ± 0.260	5.827 ± 1.064	-0.276 ± 0.165	0.572 ± 0.220
$2B_t$	0.647 ± 0.078	0.827 ± 0.096	3.910 ± 1.230	4.187 ± 2.081	4.870 ± 0.348	-0.128 ± 0.161	0.373 ± 0.081
$3B_t$	0.675 ± 0.064	0.856 ± 0.071	2.660 ± 0.547	3.320 ± 0.333	5.147 ± 0.316	-0.242 ± 0.080	0.401 ± 0.020
$4B_t$	0.235 ± 0.011	0.298 ± 0.014	1.250 ± 0.089	1.590 ± 0.560	1.850 ± 0.062	0.265 ± 0.249	0.048 ± 0.0098
$5B_t$	0.359 ± 0.065	0.463 ± 0.088	2.020 ± 0.754	2.020 ± 0.372	2.837 ± 0.721	-0.326 ± 0.157	0.101 ± 0.030
$6B_t$	0.518 ± 0.047	0.644 ± 0.055	2.123 ± 0.320	2.553 ± 0.170	3.757 ± 0.399	-0.026 ± 0.012	0.228 ± 0.065
$7B_t$	0.602 ± 0.076	0.757 ± 0.099	2.917 ± 0.815	3.133 ± 0.743	4.413 ± 0.433	-0.095 ± 0.101	0.342 ± 0.090
$8B_t$	0.933 ± 0.905	1.180 ± 0.141	5.690 ± 0.212	4.460 ± 1.343	6.580 ± 0.877	-0.044 ± 0.106	0.724 ± 0.253
$9B_t$	0.528 ± 0.024	0.672 ± 0.021	2.335 ± 0.007	2.605 ± 0.672	4.130 ± 0.113	0.166 ± 0.094	0.261 ± 0.008
$10B_t$	0.384 ± 0.064	0.476 ± 0.081	2.815 ± 1.930	1.630 ± 0.198	2.695 ± 0.233	0.061 ± 0.002	0.103 ± 0.016
$11B_t$	0.545 ± 0.075	0.700 ± 0.110	2.485 ± 0.587	2.645 ± 0.601	4.610 ± 1.103	-0.023 ± 0.063	0.368 ± 0.170
$12B_t$	0.466 ± 0.057	0.588 ± 0.064	2.085 ± 0.049	2.295 ± 0.163	3.695 ± 0.092	-0.006 ± 0.099	0.214 ± 0.008
$13B_t$	0.113 ± 0.008	0.147 ± 0.001	0.739 ± 0.397	0.518 ± 0.001	0.955 ± 0.134	0.444 ± 0.668	0.018 ± 0.001
$14B_t$	0.616 ± 0.083	0.786 ± 0.121	3.010 ± 0.764	3.275 ± 0.870	4.605 ± 1.011	0.018 ± 0.274	0.336 ± 0.085

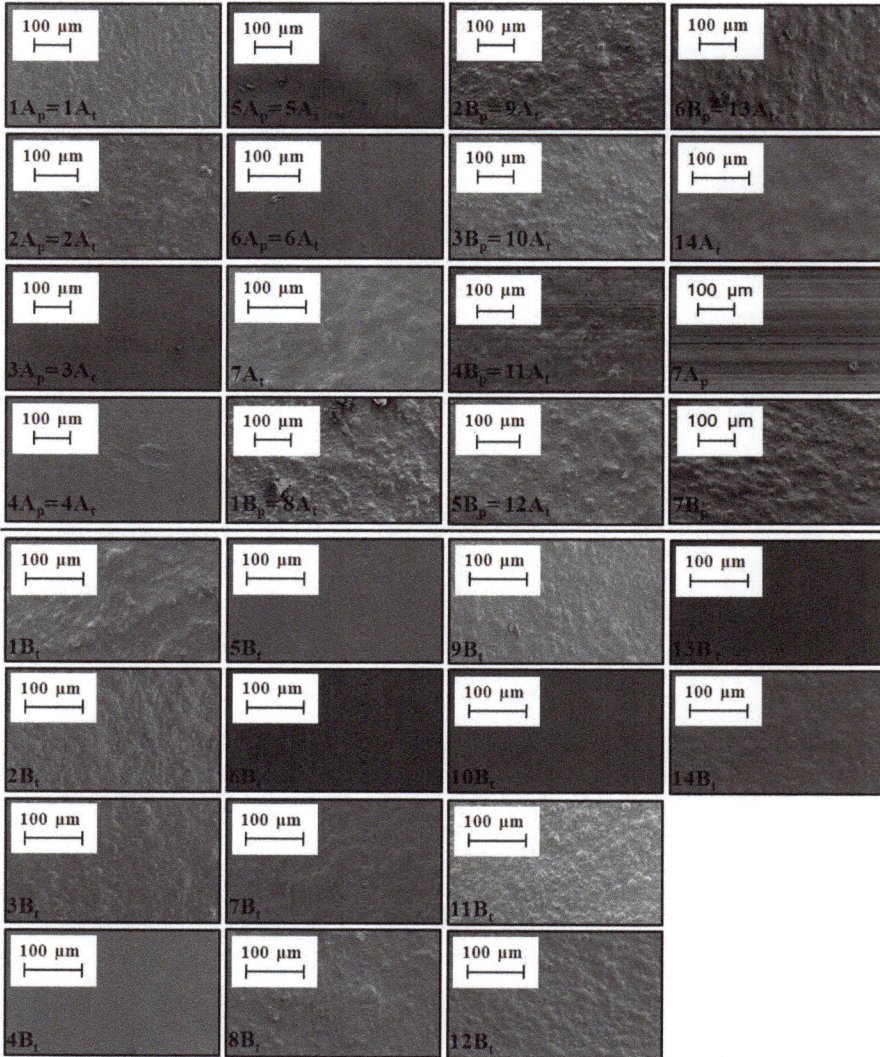

Figure 6.1 FESEM microscopies of the tested PS surfaces.

A

B

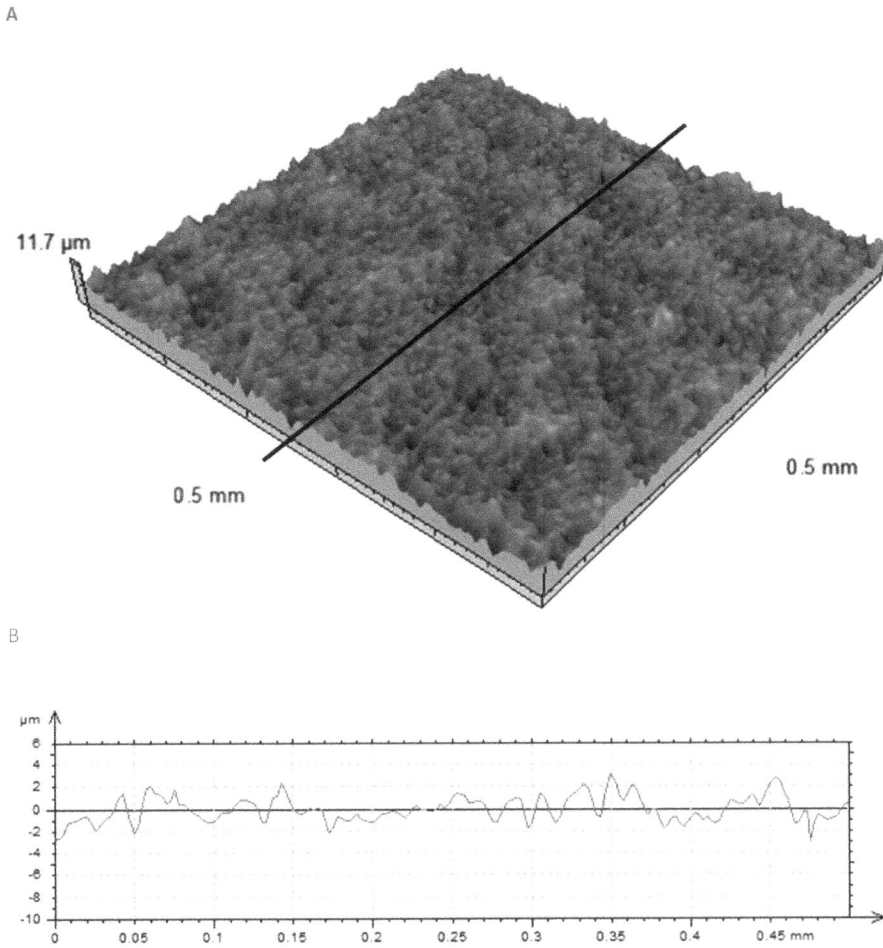

Figure 6.2 Surface topography before plasma treatment. PS surface of sample 2A$_p$ as a representative of surface topography of samples 1A$_p$ and 2A$_p$. (A) 3D topography and (B) 2D profile (extracted at 50 mm from the edge of the square measured area).

A

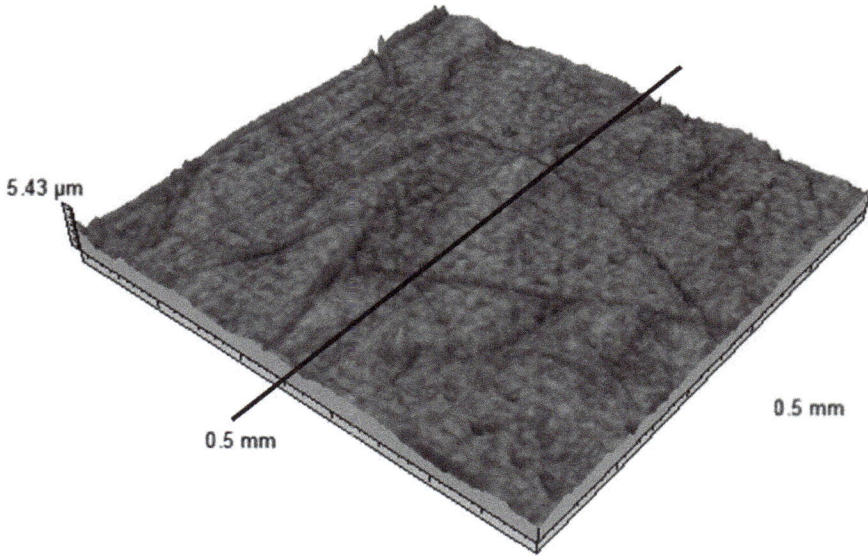

B

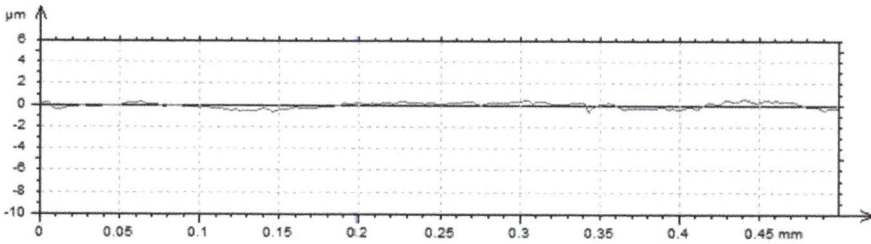

Figure 6.3 Surface topography before plasma treatment. PS surface of sample $3A_p$ as a
representative of surface topography of samples $3A_p$, $4A_p$, $6A_p$, and $7A_p$. (A) 3D topography and (B) 2D
profile (extracted at 50 mm from the edge of the square measured area).

A

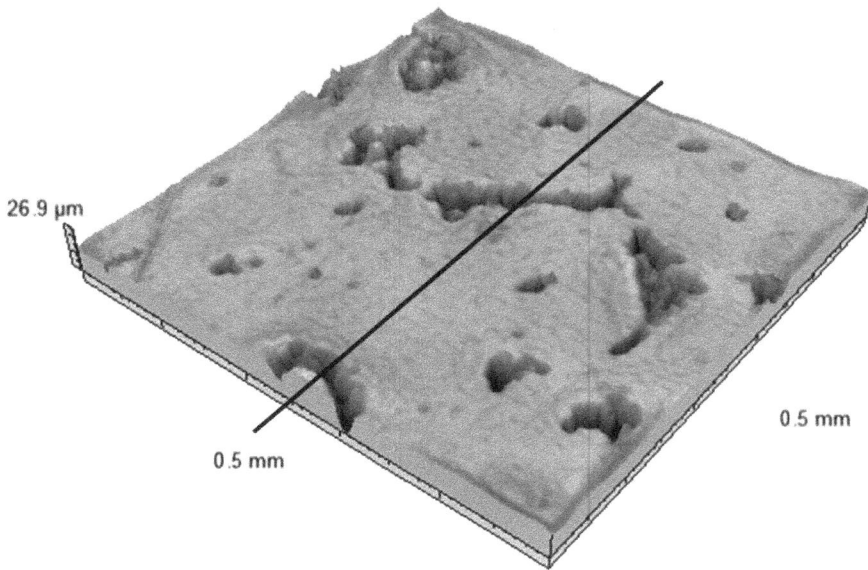

B

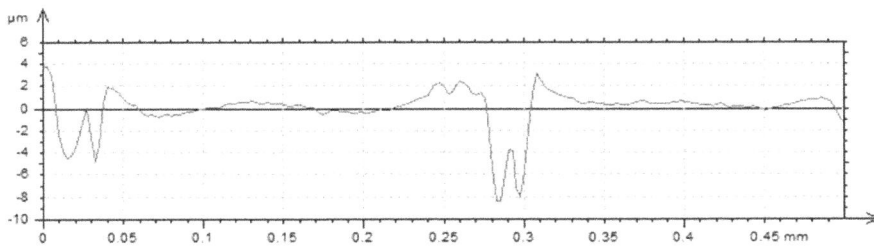

Figure 6.4 Surface topography before plasma treatment. PS surface of sample 5A$_p$. (A) 3D topography and (B) 2D profile (extracted at 50 mm from the edge of the square measured area).

A

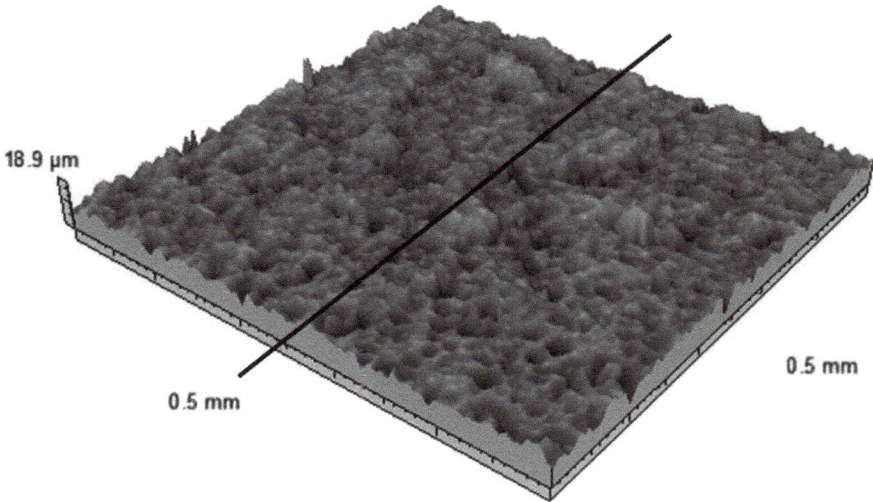

B

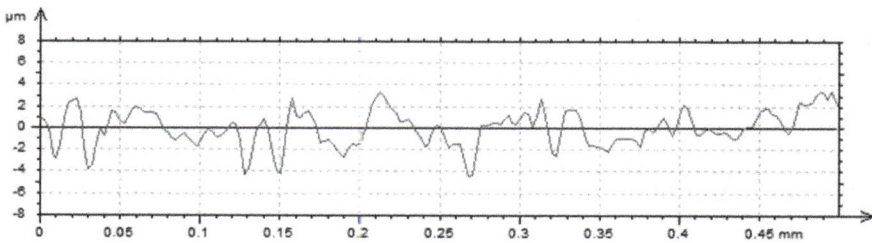

Figure 6.5 Surface topography after plasma treatment. PS surface of sample 4B$_p$ as a representative of surface topography of plasma treated samples. (A) 3D topography and (B) 2D profile (extracted at 50 mm from the edge of the square measured area).

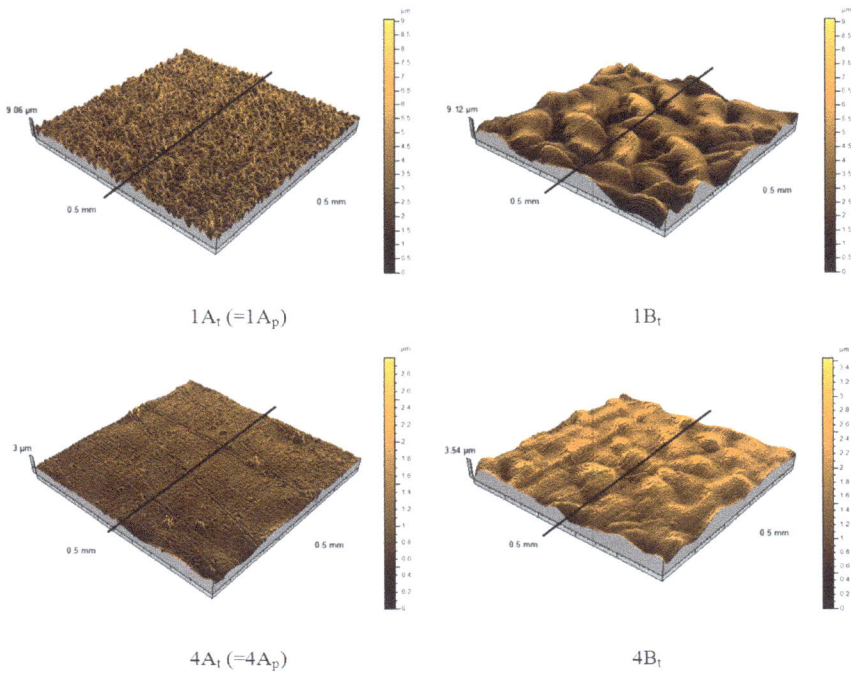

$1A_t (=1A_p)$

$1B_t$

$4A_t (=4A_p)$

$4B_t$

Figure 6.6 3D PS surface topography of sample 1 (up) and 4 (down), before (left) and after (right) thermoforming treatment.

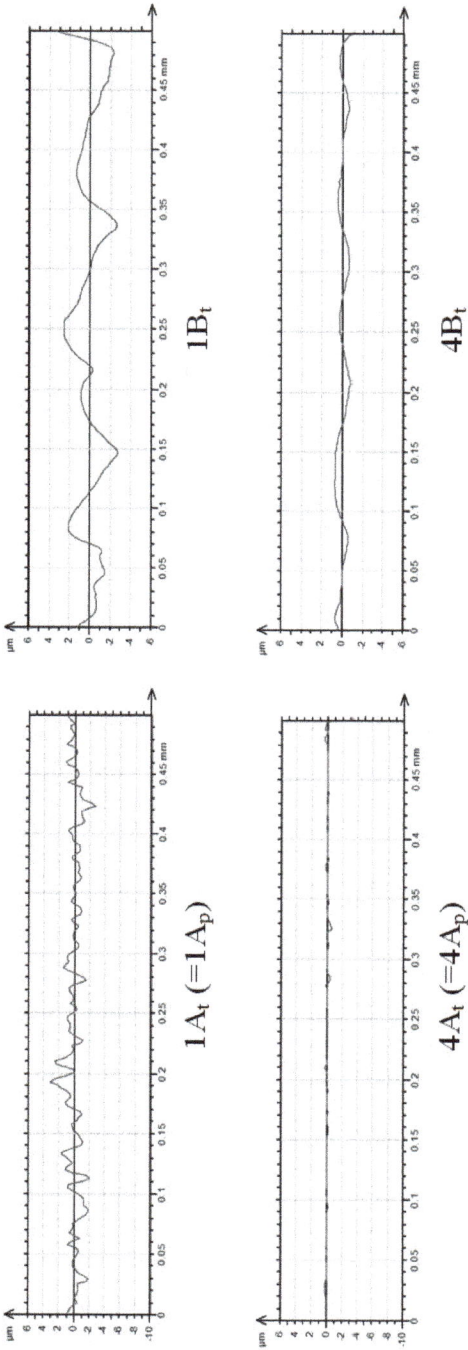

Figure 6.7 2D PS profiles of sample 1 (up) and 4 (down), before (left) and after (right) thermoforming treatment. Each profile was extracted at 50 mm from the edge of the square measured area shown in Fig.6.6.

Figure 6.8 FESEM microscopies of the lotus (*Nelumbo nucifera*) leaf: a natural 6-month dried adaxial leaf surface (a and b), the papillose cells (c), and the wax tubules (d).

Figure 6.9 Sample 4B, at 36°; sliding was observed at 48°.

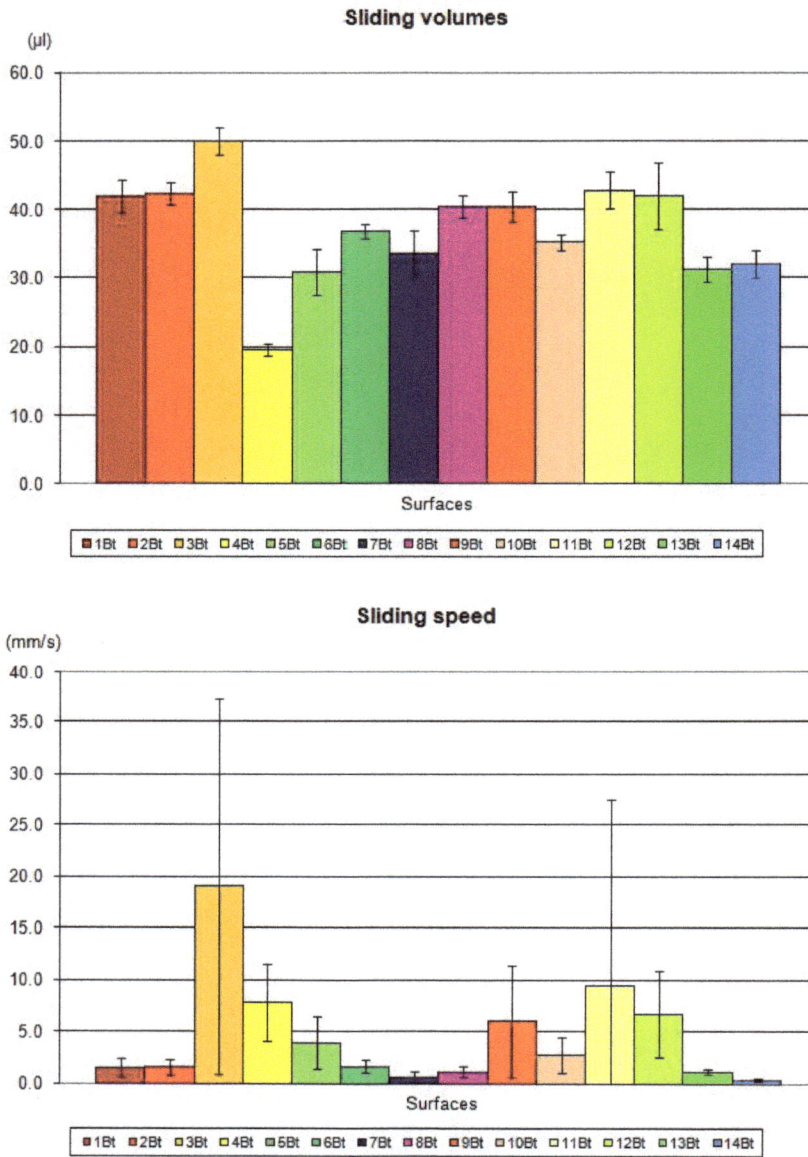

Figure 6.10 Sliding volume or speed of B, surfaces.

Table 6.2 CA measurements of the tested PS surfaces.

CA (°)		CA (°)	
$1A_p = 1A_t$	55 ± 3.2	$1B_t$	93 ± 2.5
$2A_p = 2A_t$	80 ± 5.8	$2B_t$	87 ± 3.7
$3A_p = 3A_t$	72 ± 6.7	$3B_t$	81 ± 1.9
$4A_p = 4A_t$	78 ± 7.6	$4B_t$	91 ± 4.2
$5A_p = 5A_t$	69 ± 4.0	$5B_t$	82 ± 2.3
$6A_p = 6A_t$	88 ± 3.8	$6B_t$	94 ± 2.8
$7A_t$	89 ± 2.4	$7B_t$	88 ± 3.2
$1B_p = 8A_t$	50 ± 6.7	$8B_t$	82 ± 2.4
$2B_p = 9A_t$	84 ± 4.4	$9B_t$	89 ± 3.0
$3B_p = 10A_t$	67 ± 3.3	$10B_t$	85 ± 5.0
$4B_p = 11A_t$	50 ± 7.1	$11B_t$	84 ± 2.8
$5B_p = 12A_t$	61 ± 6.1	$12B_t$	78 ± 4.4
$6B_p = 13A_t$	87 ± 6.4	$13B_t$	77 ± 5.3
$14A_t$	81 ± 2.8	$14B_t$	82 ± 1.9
$7A_p$	78 ± 5.4		
$7B_p$	83 ± 5.0		

6.4. Discussion

6.4.1. Plasma Treatment

According to Wenzel: $\cos\theta_A = r_{A.B} \cdot \cos\theta_B$, where $r_{A.B} = \dfrac{r_B}{r_A}$, r_A (1.0006 - 1.0558) and r_B (1.0629 - 1.2080) are the Wenzel roughness parameters (reported in Table 6.3) before and after the plasma treatment respectively. $\theta_{A.B}$ is the corresponding theoretical contact angle. Thus the effect of plasma treatment can be evaluated by the increment of superficial roughness. The comparison between theoretical predictions and experimental data is presented in Fig.6.11.

According to the FESEM microscopies reported in Fig.6.1, the plasma treatment increases surface roughness. It is necessary to consider sample $5A_p$ separately as it presents a specific initial (untreated, Fig.6.4) condition showing several distributed valleys with greater depth than those observed in other samples. This condition shows the greatest value of the *Sdr* parameter (11 %); after plasma treatment, the *Sdr* parameter is of the same order of magnitude as the other samples (Table 6.1). The plasma treatment levels surfaces with deep valleys, as seen in sample 5, and eliminates the presence of excessive high peaks by surface erosion. With the exception of sample 5, the plasma treatment increases the roughness parameters (see *Sa, Sq, Sp, Sv, Sz* in Table 6.1), leading to more valleys than peaks (negative value of *Ssk*) with a greater effective area than that of the untreated surfaces (greater value of *Sdr*). Apart from samples $2A_p$ and $7A_p$, we observe a decrement of CA as expected from the Wenzel theory for an intrinsically hydrophilic material subjected to an increment of roughness. Thus plasma treatment is ideal for increasing PS surface wettability.

Table 6.3 Wenzel roughness parameters *r* of PS surfaces.

	$1A_p$	$2A_p$	$3A_p$	$4A_p$	$5A_p$	$6A_p$	$7A_p$
r_A	1.0558	1.0528	1.0027	1.0011	1.1100	1.0006	1.0063
	$1B_p$	$2B_p$	$3B_p$	$4B_p$	$5B_p$	$6B_p$	$7B_p$
r_B	1.2080	1.1450	1.0709	1.1297	1.0629	1.1166	1.1553

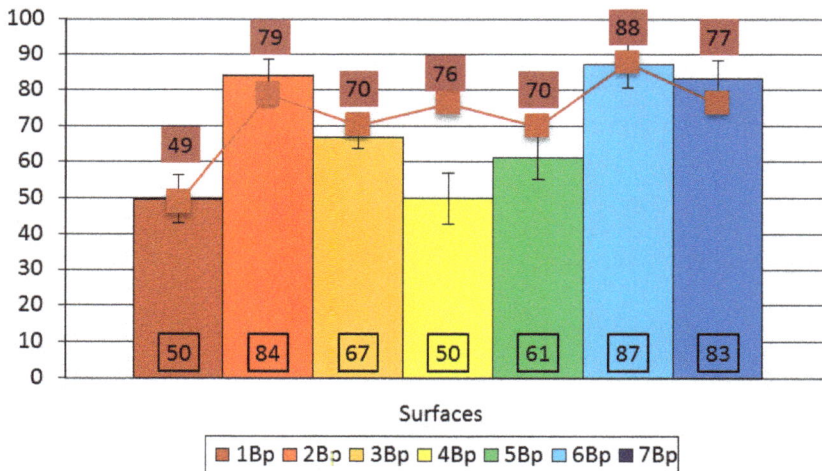

Figure 6.11 Experimental measurements vs. theoretical predictions of CA for samples after plasma treatment.

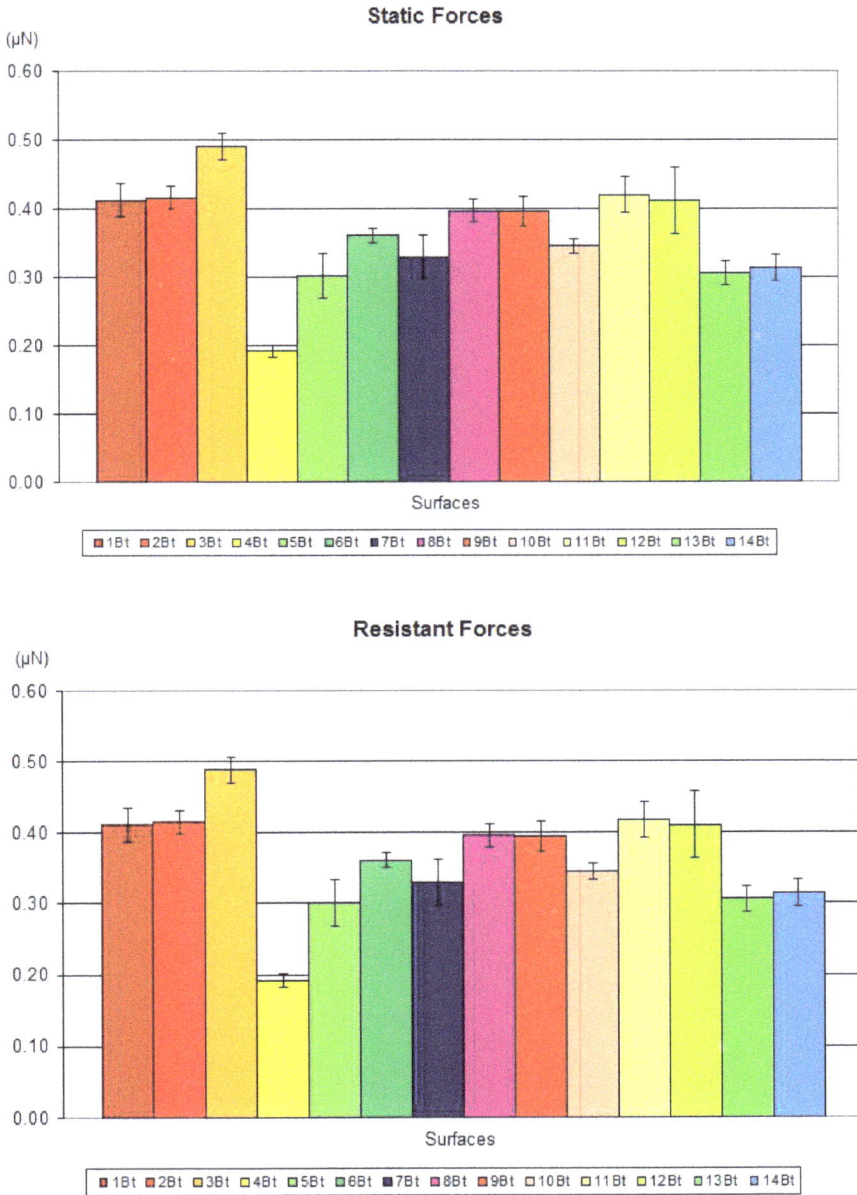

Figure 6.12 Static and resistant forces on B, surfaces.

6.4.2. Thermoforming Treatment: Adhesive Static and Resistant Forces

Considering the roughness parameters reported in Table 6.1 and the profilometer 3D-images of Fig.6.6, we observe that the thermoforming treatment globally decreases the roughness parameters (see Sa, Sq, Sp, Sv, Sz in Table 6.1). The Sdr parameter close to 0 % suggests that the thermoforming treatment smooths the surfaces. Apart from sample 13A$_t$, we observe an increment of the CA as expected from the Wenzel theory for an intrinsically hydrophilic material subjected to a decrement of the roughness. Finally, we calculate the static and the resistant forces of sliding droplets for B$_t$ vertical (at 90°) surfaces compared with those on a natural 6-month dried lotus leaf. The complete measured wettability parameters for the lotus leaf are summarized in Table 6.4.

Table 6.4 Contact angle, sliding angle, sliding volume, and sliding speed of a natural 6-month dried adaxial leaf lotus surface.

	Lotus (*Nelumbo nucifera*)
CA (°)	153.4 ± 3.28
TA (°)	26.2 ± 3.64
SV (µl)	4.7 ± 1.15
SS (mm/s)	233.3 ± 25.82
Static force (µN)	0.043 ± 0.008
Resistant force (µN)	0.032 ± 0.009

The static force (F_S) is computed as follows:

$$F_S = m \cdot g = V \cdot \rho_0 \cdot g$$

(1)

where V is the drop sliding volume, ρ_0 is the water density, and g is the gravity acceleration. The resistant force (F_R) is obtained assuming a resistant force during sliding on PS is proportional to the low velocity observed, as:

$$F_R = F_S \cdot \left(1 - \frac{F_{RL}}{F_S} \frac{v}{v_{oL}}\right)$$

(2)

where F_S is the static force of the surface; F_{RL} and v_{oL} are the resistant force (0.032 ± 0.009 µN) and the sliding speed (233 ± 25.82 mm/s) for the lotus leaf, respectively; v is the sliding speed of the surface. The resistant force of the lotus leaf was computed as proportional to the square of velocity, due to the high velocity observed:

$$F_{RL} = \frac{1}{2} \cdot \rho_0 \cdot v_{oL}^2 \cdot A_r \cdot C_p$$

(3)

where A_r is the resistant area (2.32 ± 0.327 mm²) and C_p is the drag coefficient (estimated to be ~0.47 since the shape of the sliding droplet is nearly a sphere). Thus, $F_{RL} \approx 0.03$ μN. The resistance forces are considered negligible, and thus static and resistant forces are nearly identical (Fig.6.12).

6.5. Conclusions

In this study, the effects of plasma and thermoforming treatments on water sliding behaviour have been studied on fourteen different PS surfaces in terms of contact angle, sliding angle, sliding volume, sliding speed, and static and resistant forces acting on the sliding droplet. These experimental results are compared with those on a natural 6-month dried lotus leaf. A significant correlation between technological treatment, surface roughness parameters, and wetting measurements emerges. The present analysis suggests that plasma/ thermoforming are ideal treatments to tune the wettability and enhance the hydrophilic/hydrophobic behaviour of PS surfaces.

Chapter **7**

A Superhydrophobic Polystyrene by Replicating the Natural Lotus Leaf

Abstract

In this study, we report the realization of an artificial biomimetic superhydrophobic polystyrene (PS) surface via direct copy of a natural lotus leaf using a simple template method at room temperature and atmospheric pressure. The water sliding behaviour was characterized by measuring the contact angle (CA), sliding angle, sliding volume and sliding speed (SS) of the lotus leaf (CA = 153.4°, SS = 319.4 mm/s), copied lotus leaf, negative silicone template, flat silicone, PS control surfaces, and final PS artificial leaf (CA = 149.0°, SS = 416.7 mm/s). The PS artificial leaf displays properties comparable with those of lotus leaves. This template method needs neither expensive instruments nor complicated chemical treatments. An adequate optimization of this molding process into automated industrial procedures may lead to a new, innovative, cheap concept for large-scale industrial development of superhydrophobic surfaces, starting from intrinsically hydrophilic counterparts, as demonstrated here for PS.

7.1. Introduction

The ability of some natural leaves to stay un-wetted (superhydrophobic) and dirt-free (self-cleaning) was observed more than 2,000 years ago; however, only in the twentieth century have scientists studied these two related phenomena on natural leaves, focusing on their natural morphologies (Barthlott, 1981; Neinhuis, 1997; Barthlott, 1997) correlated with surface roughness (Herminghaus, 2000; Wagner, 2003; Nosonovsky, 2005; Lu-quan, 2007; Otten, 2004; Zhiqing, 2007; Bhushan, 2007a; Nosonovsky, 2006; Nosonovsky, 2007a; Koch, 2009), surface adhesion (Lepore, 2008; Pugno, 2008a, Pugno, 2008b), friction (Bhushan, 2008), and self-cleaning (Quéré, 2002). The most famous is the lotus (*Nelumbo nucifera*), on which "raindrops take a clear, spherical shape without spreading, which probably has to be ascribed to some kind of evaporated essence", as Goethe described in 1817 (Solga, 2009). Superhydrophobicity and self-cleaning are said to be correlated, but this correlation do not always appear in nature. For example, the water ferns of *Salvinia* possess trichomes and waxes, as surface micro- and nano-structures, which facilitate superhydrophobicity but not self-cleaning (Koch, 2009).

After better comprehension of these natural properties, it is desired to implement them in man-made technology. This process is well-known as biomimicry, from the Greek word *biomimesis*, which means "mimic life". In fact, micro-, nano-, and micro/nano-patterned superhydrophobic surfaces have become one of the most popular research topics in engineering (Lee, 2008b; Raibeck, 2008). Due to the superhydrophobicity and self-cleaning characteristics of its surfaces, the natural lotus leaf has been extensively bio-mimed (Lee, 2008b; Sun, 2005b; Furstner, 2005; Lee, 2006). A number of methods have been applied to fabricate artificial surfaces mimicking its natural morphology: the fabrication of nano- (micro-) protrusion by reactive ion etching (Lee, 2008b); the creation of structured coatings similar to lotus leaves from polyelectrolyte multilayer films (Zhai, 2004); construction of a nickel mold *via* electroforming and UV-nanoimprint lithography (Lee, 2006a; Lee, 2007; Lee, 2006b); the addition of ethanol to PS solution (Yuan, 2007); the production of Poly(dimethyl siloxane) (PDMS) replica molding with photolithographically manufactured micro-patterned masters (Yeo, 2010); nano-casting PDMS (Sun, 2005b); by soft-lithography method of Poly(methyl meth-acrylate) (PMMA) replica molding using PDMS (Singh, 2007); the development of a dental wax cast technique (Gorb, 2007) of a replica with polyether (PE) (Furstner, 2005), Poly(vinyl siloxane) (Koch, 2007), conventional lacquer (Wagner, 2003), and epoxy resin (Schulte, 2009; Cheng, 2006); artificial surface patterning (Oner, 2000; Shibuichi, 1996; Miwa, 2000; Jung, 2007a; Jung, 2007b; Jung, 2008; Bhushan, 2007b; Nosonovsky, 2007a; Nosonovsky, 2007b; Nosonovsky, 2007c; Nosonovsky, 2007d; Nosonovsky, 2008); chemical surface modification (Coulson, 2000; Hozumi, 1998; Erbil, 2003); a combination of both morphological and chemical modifications (Burton, 2005; Feng, 2002; Lau, 2003). However, only six patents for invention, with "self cleaning super hydrophobic" as keywords, have been awarded during the last 6 years at the European Patent Office, thus at an average rate of one European patent per year.

In the literature, superhydrophobic and self-cleaning properties have been evaluated only by measuring the contact angle (CA) and the contact angle hysteresis (CAH), which can be more easily quantified by the tilting angle (TA) (Bhushan, 2008; Sun, 2005b; Yuan, 2007; Yeo, 2010). By definition, a high CA and low CAH (or TA) denote a superhydrophobic and self-cleaning surface (Koch, 2009; Bhushan, 2008; Sun, 2005b; Lee, 2006b). However, it was highlighted that, more than the maximum CA, the TA is the important parameter to determine if a surface is superhydrophobic, due to its correlation with the driving force of a liquid droplet (Zhiqing, 2009). In a previously published work from our group (Lepore, 2012a), we introduce two more parameters - the sliding volume (SV) and the sliding speed (SS) - as additional indexes for the superhydrophobicity and self-cleaning properties of a surface. This addendum was motivated by the fact that SV and SS are straightforward and direct measurements of the surface water-repellency and self-cleaning ability, in both static and dynamic regimes.

Moreover, we have defined a droplet minimum volume and the corresponding sliding speed with respect to a vertical surface very close to a real condition of use (*i.e.*, glass windows, external building coverings, internal faces of refrigerators or freezers, surfaces of bathroom fittings or tiles, etc.). A high CA, low CAH (or TA), low SV, and high SS denote a superhydrophobic and self-cleaning surface.

Referring to previously published scientific works on lotus leaves, only a few papers including the present study discuss the replication of its surface structures (convex cell papillae but not 3D wax crystals) by molding (Sun, 2005b; Lee, 2006a; Lee, 2007; Lee, 2006b).

In this study, we readily obtain a biomimetic lotus-leaf-like polystyrene (PS) superhydrophobic surface by replicating the morphological surface pattern of a natural lotus leaf. The molding method is similar to that reported in previously published works (Zhiqing, 2007; Sun, 2005b) but, to our knowledge, our method is the first capable of creating a superhydrophobic lotus-leaf-like PS surface using a template method at room temperature and atmospheric pressure (no controlled temperature/vacuum condition was necessary). We elaborate on the definition of a superhydrophobic and self-cleaning surface, taking into account not only the classical parameters (CA and TA) but also the SV and SS parameters of a rolling droplet, for a more complete surface characterization of a lotus leaf, copied lotus leaf, negative silicone template, flat silicone and PS control surfaces, and positive PS template. Compared with the other mentioned methods, our molding technique needs neither expensive instruments nor complicated chemical treatments and is thus a good candidate for industrial applications.

7.2. Materials and Methods

7.2.1 Molding Method

Upper leaf sides (adaxial) of fresh lotus plant, cultivated in the "Giardino Botanico Rea" (Turin) associated with the Natural Science Museum of Turin, were used. The leaves (diameter of ~25 cm) were cut and the first copy (C1) deposition was made within 24 h. Fig.7.1 describes a simple flowchart of the lotus leaf replication process performed in two steps at room temperature and atmospheric pressure.

The molding method uses a silicone elastomer (R39-2186-2, Nusil Technology) amd a low-viscosity hydrophobic silicone to obtain the first copy C1 (diameter of ~25 cm) of a natural lotus leaf template (LL). The lotus leaf copy is named copied lotus leaf (CLL). R39-2186-2 is a two-phase silicone, mixed in mass proportion of 1:1, which is extracted from side-by-side kits through a

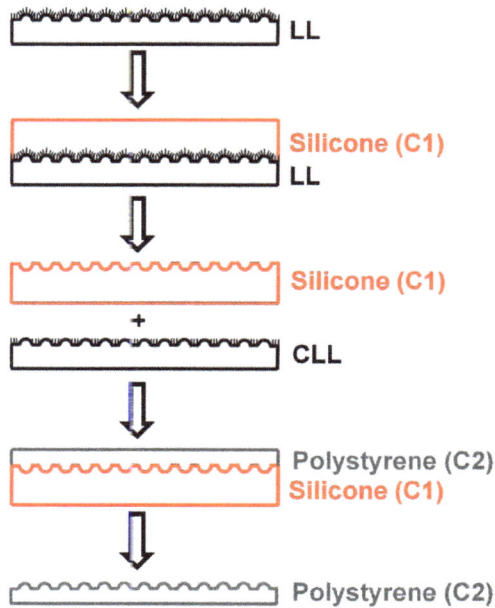

Figure 7.1 Schematic illustration of the presented lotus leaf replication process, which is composed of two steps at room temperature and atmospheric pressure.

disposable static mix tip. Both components are extruded directly onto the lotus leaf and immediately spread with a stick to form a few millimetres-high silicone layer on the substratum. After polymerization, the negative mold can easily be peeled off from the surface, giving rise to C1. Without any other intermediate treatment, C1 was directly used for preparation of the positive mold, C2. The low viscosity of the silicone R39-2136-2 does not require any pressure to replicate smaller structures on the leaves unlike the molding methods already described in the literature (Lee, 2006a; Lee, 2007; Lee, 2006b; Koch, 2008).

A commercial hydrophilic PS sheet was reduced into small particles without any further treatment. A volume of 20 ml tetrahydrofuran solvent was added to 1 g of PS, and the solution was stirred with a magnetic stirrer (Are - Velp) for 20 min at increasing speed (5 min at 600 rpm, 5 min at 720 rpm and 10 min at 840 rpm) to form a uniform solution at room temperature and pressure. The solution was directly cast on a 9-cm diameter subarea of the negative silicone template (C1). After solvent volatilization for 24 h at atmospheric pressure and room temperature, a double adhesive was applied on the rigid substratum. The PS positive template (C2) was attached on the double adhesive, and the silicone negative template was then peeled off from the C2 positive (diameter of ~9 cm). The surface micro-structure of the lotus leaf was transferred in this way to the PS surface on the side contacting the silicone.

Two control surfaces have been characterized to establish the reference intrinsic parameters for comparison between C1 and C2: R39-2186-2 silicone and the PS/20ml-tetrahydrofuran solution was cast on a clean silicon wafer in 100 % ethanol sonicated, and allowed to polymerize (volatilize) for 24 h, which provided flat silicone and PS surfaces, called C1_control and C2_control.

7.2.2. Surface Characterization

We observed the surfaces of LL, CLL, C1, and C2 by means of a field emission scanning electron microscope (FESEM, Zeiss SUPRA 40 for LL and CLL and FEI-Inspect™ F50 for C1 and C2) equipped with a field emission tungsten cathode. Samples of ~0.5 cm² were obtained, fixed to aluminium stubs by double-sided adhesive carbon conductive tape (Nisshin EM Co. Ltd.), used as these were (except for C1, which was cleaned with ethanol) and air-dried. Samples LL and CLL (C1 and C2) were Cr(Au-Pd) coated, approximately 10 nm in thickness. No fixation processes were applied to LL and CLL to avoid alteration of the wax crystals (Neinhuis, 1997).

7.2.3. Wettability Measurement

The wettability of LL, CLL, C1, C1_control, C2, and C2_control surfaces was determined by measuring the static CA of distilled water droplets over samples fixed to a horizontal plane by a soft adhesive. We consider a series of 20 (five were previously considered (Lee, 2008b; Yeo, 2010; Lepore, 2012)) random-volume droplets gently deposited on LL and CLL (C1, C1_control, C2, C2_control) with a standard single-use syringe. The contact angle was recorded with an OLYMPUS MJU 1010 digital photocamera, measured, and statistically analysed with the software ImageJ 1.41o. The average CA of the control surfaces was used as the intrinsic CAs of the R39-2186-2 and PS flat surfaces.

Two conceptually distinct procedures were used to evaluate droplet sliding characteristics: (1) fixing the volume of the droplet (~18 µl, the diameter of the spherical droplet was ~2.2 mm) and measuring the tilted angle of the sample stage at sliding (TA) or (2) fixing the angle of the specimen stage vertically (90°) and measuring the minimum SV of the droplet, increasing it in 2 µl volume increments. For the second procedure, the sliding speed (SS) of the droplet was also determined, measuring the time of the minimum SV droplet to cover a fixed distance of 10 mm. Fig.7.2 schematically shows the step-by-step process to determine the two additional parameters, SV and SS.

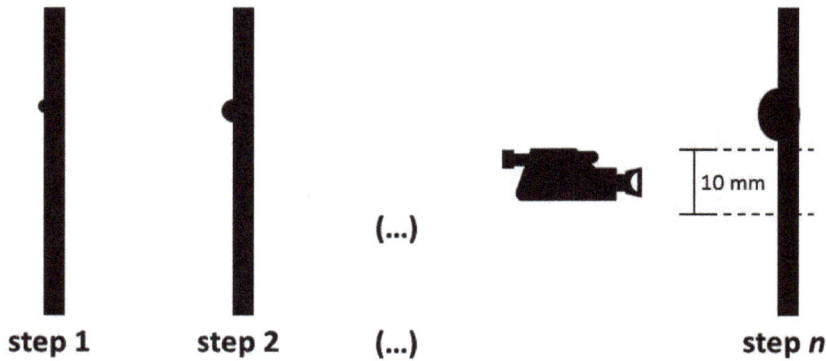

Figure 7.2 The step-by-step process to determine the two additional parameters, SV and SS. The specimen stage was fixed vertically (90°), and the droplet volume was increased in 2-μl increments, from 2-μl up to the minimum sliding volume (SV) of the droplet, at which sliding occurs (final step, *n*). The sliding speed (SS) was determined by measuring the time required to cover a fixed distance of 10 mm (mean velocity).

7.3. Results

7.3.1. Surface Characterization

This technique demonstrates excellent replication for convex micro-structures (cell papillae) of the lotus leaf and high replication quality of these micro-structures from LL to C1 and in turn to C2. As shown in Fig.7.3 (with the lack of 200-nm-scale bar micrographies for C1 and C2), the nano-tubules (superimposed layer of hydrophobic 3D wax tubules) of the lotus leaf have not been transferred because of their permanent removal during C1 deposition. This prevents nano-tubule replication, as reported in previously published studies (Wagner, 2003; Furstner, 2005; Lee, 2006a; Lee, 2007; Lee, 2006b; Singh, 2007; Koch, 2008).

7.3.2. Wettability Measurement

In Table 7.1, mean values and standard deviation of wettability measurements are reported; correspondingly, Fig.7.3 shows a representative water droplet on each surface. The first remarkable result concerns the same CA (~150°) for LL and CLL despite the absence of wax nano-tubules over the cell papillae on CLL. When compared to the flat C1_control surface (~100°, intrinsically hydrophobic) and C2_control surface (~85°, intrinsically hydrophilic), the water

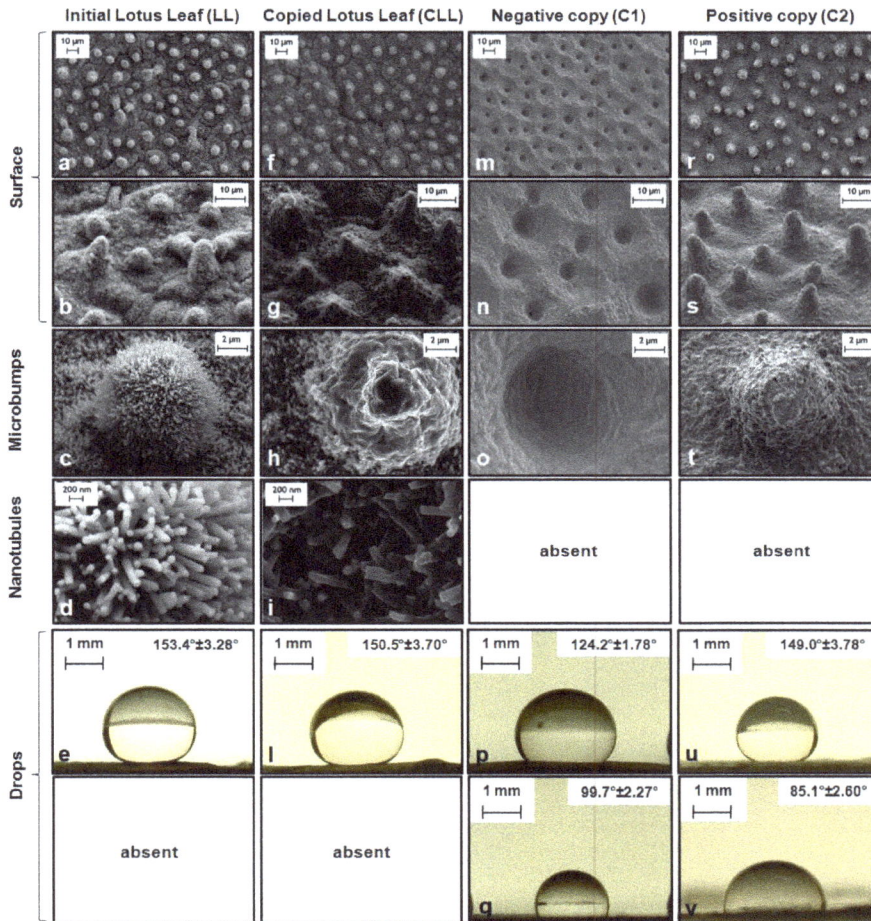

Figure 7.3 Details of: (a) fresh lotus leaf (LL), (f) lotus leaf resulted after copying process (CLL), (m) negative copy (C1), and (r) positive copy (C2). In particular, b, g, and s (n) show randomly distributed convex (concave) cell papillae; c, h, and t (o) show magnified detail of the convex (concave) cell papillae; wax tubules are magnified in d (natural wax tubules) and i (the wax tubules are broken due to C1 deposition and peeling). The nano-tubules are absent on C1 and C2. Water droplet on the surface of: (e) fresh lotus leaf (LL), (l) lotus leaf resulted after copying process (CLL), (p) negative copy (C1), and (u) positive copy (C2). (q) and (v) show the shape of a water droplet on C1_control and C2_control surfaces, respectively. For LL and CLL, no control surface can be defined. The measurements reported in e, l, p, u, q, v are the average CA ± st.dev..

CAs of corresponding replicas are increased by about 24° (C1 ~ 124°) and 64° (C2 ~ 149°). This finding indicates that the CA of C2 is comparable with LL, which suggests an excellent CA replication. The C2_control surface of PS has an intrinsic CA close to values found in literature (~92° (Bhushan, 2008), ~98° (Yuan, 2007)), while C2 has a significantly higher CA of 45° than that previously reported (~105°, Bhushan, 2007). On the other hand, for a PE-based (intrinsic CA = 102.9° ± 4.5) molding method, the CA for C2 is comparable with that of the PE lotus replica (CA = 157.8° ± 4.2) (Furstner, 2005).

The best (lowest) TA was observed for CLL, showing 10° lower than the value recorded for LL. The silicone samples, both flat and micro-structured, display a high TA (~74° and ~81°, respectively), and C2_control displays an intermediate value of ~50°. The SV is similar among LL (~5 µl) and CLL (~6 µl) surfaces whereas it increases to ~20 µl for the others samples. The worst result was observed for C2 (TA not observed, thus > 90°).

Despite this, the SS highlights the accurate replication obtained for C2, which is comparable with the surfaces LL and CLL. The SS values for C2_control, C1, and C1_control are three, two, and one order of magnitude lower than those of C2, LL and CLL.

Table 7.1 Contact angle (CA), tilting angle (TA), sliding volume (SV), and sliding speed (SS) of lotus leaf (LL), copied lotus leaf (CLL), first silicone copy (C1), flat silicone control surface (C1_control), second PS copy (C2), and flat PS control surface (C2_control).

	CA (°)	TA (°)	SV (µl)	SS (mm/s)
LL	153.4 ± 3.28	26.2 ± 3.64	4.7 ± 1.15	319.4 ± 97.42
CLL	150.5 ± 3.70	18.0 ± 1.52	6.3 ± 0.82	319.4 ± 97.42
C1	124.2 ± 1.78	80.7 ± 1.32	19.3 ± 0.77	9.7 ± 2.95
C1_control	99.7 ± 2.27	73.9 ± 4.21	21.7 ± 3.44	15.7 ± 12.27
C2	149.0 ± 3.78	> 90°	23.0 ± 1.10	416.7 ± 91.29
C2_control	85.1 ± 2.60	48.6 ± 3.30	20.0 ± 0.00	0.1 ± 0.08

7.4. Discussion

The described molding technique requires a silicone polymerization time of one order of magnitude longer than a previously described fast and similar molding process (Koch, 2008). There is little discernible difference between living and dried lotus leaves (cell papillae are taller on the fresh lotus leaf) in wettability response (Bhushan, 2008; Singh, 2007; Cheng, 2006), but a longer polymerization time is needed for an excellent morphological replication. The resolution can be quantified to ~1 μm; thus this method cannot copy nano-structures with the considered materials (silicone and PS).

The material of the positive (negative) template is intrinsically hydrophilic (hydrophobic), whereas according to the experimental increment of CA, the C2 (C1) replica becomes more hydrophobic (more hydrophobic).

Previously published studies report that hydrophobic behaviour can be achieved from hydrophilic material by increasing the surface roughness (Lee, 2006a; Lee, 2006b). A possible explanation is related to the molding process induction of modifications to topological characteristics, namely the cell papillae of PS replicas. Several studies (Nosonovsky, 2005; Koch, 2009) present evidence that a surface with hemispherical topped asperities, like lotus convex cell papillae, is the most appropriate to obtain an increased CA. On the contrary, according to the classical Wenzel model (Wenzel, 1936), surface roughness increases the hydrophilicity (hydrophobicity) of an intrinsically hydrophilic (hydrophobic) material. Consequently, the Wenzel model is not directly applicable to the PS surface even though it agrees with the observations on C1 surfaces.

According to a previous study (Lee, 2006), a surface is self-cleaning if it has a very high CA and very low CAH, which is usually associated with the Cassie-Baxter (Cassie, 1944) regime. We observed on C2 samples that the just-deposited droplet remains in a state not conformal to the topology of the substratum in accordance with the Cassie-Baxter hypothesis. However, after only a few seconds, the droplet conforms to the surface topology and thus falls into the Wenzel state, displaying a high CAH/TA. Experimentally, we observe the transition from Cassie-Baxter to Wenzel state for the C2 surface (intrinsically hydrophilic, high CA, high TA) over time or due to any external disturbance to the specimen stage (Lee, 2006) (inducing evaporation and thus pressure increment) or when depositing the droplet from some height (He, 2003). Thus, it seems that in some cases (Lee, 2006a; Lee, 2007; Lee, 2006b; Liu, 2006; Cheng, 2005; Wu, 2005), including the surfaces presented in this work, the water droplets seem to be sticky at high CA (Wenzel state).

Superficial irregularities of micro-structures are most likely present on the C2 surface, leading to unstable air pockets under the droplet which are replaced by water in a few seconds. These superficial imperfections, rather than the absence of wax crystal tubules over C2, have probably caused the significant difference between C2 and LL. We interpret the results in this way because the superposition of nano-scale wax tubules to micro-scale cell papillae on the LL is not expected to supply any significant contribution. As we have verified, CLL shows an absence of wax tubules on the cell papillae and a presence of broken and numerous wax tubules in the areas between cell papillae but still shows the same properties of LL in terms of CA, TA, SV, and SS. This finding adds information to previously published works, which highlight that the complete removal of wax tubules from the surface halve the CA value (Bhushan, 2008) whereas the annealing of wax tubules, keeping the wax composition and quantity nearly unchanged on the micro-patterned surface (this morphology is really close to our C2), diminishes the initial CA of the lotus leaf by 11 % (Cheng, 2006) and facilitates a sticky behaviour of droplets (TA > 90°, as for C2 here). However, we reach a high CA of the C2 copy even though the hierarchical nano-tubules are absent.

In accordance with previously published papers (Nosonovsky, 2005; Koch, 2009), we conclude that the presence of hemispherical micro-bumps (first hierarchical level) induces an increase in CA (for C2) and the presence of additional nano-tubules (second hierarchical level) on micro-bumps decreases the TA and SV (for CLL). Therefore, an absence of nano-tubules on C2, different from a lotus leaf, might cause the observed high TA and SV with respect to those of a natural leaf.

7.5. Conclusions

We have successfully fabricated a stable biomimetic lotus-leaf-like PS superhydrophobic surface. The CA and SS of the positive PS copy of lotus leaf are 149.0° and 416.7 mm/s, respectively, which are comparable to those of a lotus leaf (CA = 153.4° and SS = 319.4 mm/s). As shown by the existing TA and SV limitations (related to CAH), our method requires further improvement to enhance these parameters. The successful replication of nano-tubules, the necessary improvement of the molding method presented here, will facilitate an even more efficient superhydrophobic and self-cleaning surface.

Despite this, our approach is very promising for the fabrication of superhydrophobic synthetic lotus leaves from various materials working in static regime (contact angle). Compared to other methods of morphological replication of natural superhydrophobic leaves, this procedure involves ambient pressure and temperature and is technically easier, requiring neither expensive instruments nor complicated chemical treatments.

Strong Materials

Chapter 8

Evidence of the Most Stretchable Egg Sac Silk Stalk of the European Spider of the Year *Meta Menardi*

Abstract

Spider silks display strong mechanical properties, even with observed differences between species and within the same species. While many different types of silks have been tested, the mechanical properties of silk stalks taken from egg sacs of the cave spider *Meta menardi* have not yet been analyzed. *Meta menardi* has recently been chosen as the "European spider of the year 2012" from the European Society of Arachnology. Here we report a study in which silk stalks were collected directly from several caves in north-west Italy. Field emission scanning electron microscope (FESEM) images show that stalks consist of a large number of threads, each with diameter of 6.03 ± 0.58 µm. The stalks were strained at a constant rate of 2 mm/min using a tensile testing machine. The observed maximum stress, strain, and toughness modulus, defined as the area under the stress-strain curve, are 0.64 GPa, 751 %, and 130.7 MJ/m^3, respectively. To our knowledge, such a huge observed elongation has never been reported for egg sac silk stalks and suggests a significant unrolling microscopic mechanism of the macroscopic stalk that, as a continuation of the protective egg sac, is expected to be composed of very densely and randomly packed fibers.

Weibull statistics were used to analyze the results from mechanical testing, and an average value for the Weibull modulus (m) is found to be in the range of 1.5 - 1.8 with a Weibull scale parameter (σ_0) in the range of 0.33 - 0.41 GPa, showing a high coefficient of correlation (R^2 = 0.97).

8.1. Introduction

Spider silks generally display strong mechanical properties (Brunetta, 2010) and have been studied extensively during the last five decades. In particular, dragline silk is noted for its unique strength and toughness. Because of the complex structure of spider silk, large scale synthetic production remains a challenge and can only be achieved through controlled self-assembly of macromolecular components with nanoscale precision (Keten, 2010a).

Individual spiders spin 'toolkits' of seven to eight different types of silks, each of which is generated from its own discrete gland(s) and spigot(s) (Blackledge, 2006). Each type of spider silk has a unique chemical composition, molecular structure, and material properties (Blackledge, 2011). Orbwebs, for example, are composite structures built from multiple types of silks, each with its own unique molecular structure and mechanical function (Blackledge, 2011).

The most studied silk type is dragline silk, which is produced in the major ampullate gland. As the name suggests, dragline silk is used as a lifeline by most spiders moving through their environment and forms the backbone of most webs (Blackledge, 2011). Minor ampullate glands produce threads that are sometimes added to major ampullate draglines or temporary spirals of the orbweb acting as a scaffold for web construction. Aciniform glands produce silk used for wrapping prey and egg case construction, and its fibers are more stretchable and tougher than dragline silk (Rousseau, 2009). Flagelliform glands are unique to araneoid-orbweaving spiders and are used in the production of catching spiral silk. In some derived taxa (like cobweb spinning theridiids) this type of silk is used to wrap prey (Eberhard, 2010). Aggregate glands produce the glue coating on viscid capture threads and are unique to araneoid spiders, whereas piriform glands are used to cement threads to substrata as well as to form silk junctions by forming attachment disks (Blackledge, 2011).

It is widely accepted that the tubuliform (or cylindrical) glands play a major role in silk production for egg sacs (Kovoor, 1987; Foelix, 1996; Foradori, 2002; Craig, 2003), and it is likely that some spiders produce egg sac silk exclusively in these glands. Tubuliform silk is produced solely by adult orbweaving females. Egg sacs are complex, layered structures containing fibres from several different glands (Gheysens, 2005; Hajer, 2009; Vasanthavada, 2007). This complexity creates confusion about how tubuliform silk is utilized. However, the morphology of the silk is quite distinctive because the glands produce large fibers with an irregular surface that is unlike any other silk. Moreover, the left and right fibers are coated with a gluey secretion that causes them to adhere together (Gheysens, 2005). The mechanical behaviour of the silk is distinct in displaying a prominent yield followed by a long low modulus extension (Keten, 2010b; Blackledge, 2011; Van Nimmen, 2006).

In orbweb spiders, the spinnerets are three paired appendage-like organs on the abdomen, each of which contains dozens to hundreds of spigots connected to their own internal silk-producing glands (Fig.8.1) (Vehoff, 2007). A single spider is therefore capable of producing multiple silk threads of many kinds, and the arrangement of spigots on the spinnerets appears to relate functionally to how different silks are used together (Eberhard, 2010). Dragline silk, flagelliform silk, aggregate silk, and aciniform silk have been extensively characterized in *Argiope trifasciata* (Forsskål) (Platnick, 2011; Perez-Rigueiro, 2001; Elices, 2005; Poza, 2002; Guinea, 2003; Hayashi, 2004); *Araneus diadematus* (Linnaeus) (Van

Nimmen, 2005a; Madsen, 1999; Van Nimmen, 2005b; Vollrath, 2001; Gosline, 1999; Köhler, 1995; Römer, 2008; Gosline, 1986; Ortlepp 2008; Shao, 2008); *Argiope argentata* (Fabricius) (Blackledge, 2006; Swanson, 2006); *Argiope bruennichi* (Scopoli) (Zhao, 2006); *Araneus gemmoides* Chamberlin & Ivie (Swanson, 2006; Stauffer, 1994); *Larinioides* (=*Araneus*) *sericatus* Clerck (Denny, 1975); *Nephila edulis* (Labillardière) (Madsen, 1999; Vollrath, 2001); *Nephila clavipes* (Linnaeus) (Swanson, 2006; Stauffer, 1994; Dunaway, 1995; Cunniff, 1994; Vollrath, 1996); *Nephila pilipes* Fabricius (Dunaway, 1995); *Nephila madagascariensis* (= *N. inaurata madagascariensis*) (Vinson) (Gosline, 1986); *Lactrodectus hesperus* Chamberlin & Ivie (Swanson, 2006; Moore, 1999); *Leucauge venusta* Walckenaer (Swanson, 2006); *Plectreurys tristis* Simon (Swanson, 2006); *Kukulcania hibernalis* Hentz (Swanson, 2006); *Salticus scenicus* (Clerck) (Ortlepp, 2008). These studies have shown that various silk types, produced by different glands, have very different mechanical properties (Van Nimmen, 2005a; Van Nimmen, 2005b; Stauffer, 1994), giving the threads different characteristics depending on their respective function (Foelix, 1996) that may vary according to different species. Variability in the mechanical properties of spider silk is very important. Spider silk is central to many aspects of spider biology and ecology, from communication to prey capture. Spiders are the only animals which use silk in almost every part of their lives. Because of its importance, it has presumably been subjected to strong selective pressures during the 400 million years of spider evolution and can be regarded as one of the keys to spides evolutionary success (Craig, 2003; Sensenig, 2010).

It has been demonstrated that silk properties in terms of different reeling methods (Swanson, 2006; Boutry, 2011), environmental conditions (Foelix, 1996; Guinea, 2003), and types of silk (e.g. dragline, viscid or egg sac silk) (Van Nimmen, 2005a; Van Nimmen, 2005b; Stauffer, 1994) are species-specific and lead to silk-based peptide fibrils or protein aggregates with different structural and mechanical properties. For example, different reeling speeds cause a variation in the diameter of dragline thread (Vollrath, 2001) and thus the stress-strain curve varies depending on the thickness of the thread. Spider dragline silk was tested in a wet environment to show that moisture induces supercontraction for higher than 70 - 75 % humidity. When a thread is exposed to moisture, stresses quickly build up and tighten the thread (Guinea, 2003). By varying the conditions under which the spiders were kept (different reeling speeds, starvation periods), it was observed that dragline silk has different mechanical properties and varies on an interspecific, intraspecific, and intra-individual level (Madsen, 1999).

All silks are proteinaceous and belong to the general class of hierarchical protein materials. Each thread of spider silk is a composite of semi-amorphous α-chains and β-pleated nanocrystals (Keten, 2010b). In the orb web spider *Araneus diadematus* (the common European garden spider), the β-sheets consist of a series of highly conserved poly-Ala repeats and are stacked to form protein crystals; these crystals are embedded in a matrix of loosely arranged glycine-rich amino acids (Gosline, 1999). The protein crystals are held together by hydrogen bonds, one of the weakest chemical bonds, which serve an important role in defining the mechanical properties of silk. When an external force is applied, loose amino acids stretch and are straightened from a disordered position, whereas the β-sheets are subject to tensile forces (Krasnov, 2008). The β-sheet rich crystalline units are responsible for the toughness of the silk thread, and the remaining apparently amorphous regions have a rubber-like behavior (Gosline, 1984). One study uses a simple coarse-grained model to simulate the mechanical deformation of silk in which the silk constitutive unit was a combination of two domains representing α-chains and β-pleated sheets (Nova, 2010). The stress-strain curve of the simulation has a similar shape to that of silk.

The studies on dragline silk have provided an opportunity to find a natural fiber with strong tensile properties in terms of large deformation (Blackledge, 2006; Foelix, 1996; Perez-Rigueiro, 2001; Elices, 2005; Poza, 2002; Hayashi, 2004; Van Nimmen, 2005a; Madsen, 1999; Van Nimmen, 2005b; Vollrath, 2001; Gosline, 1999; Köhler, 1995; Römer, 2008; Gosline, 1986; Ortlepp, 2008; Swanson, 2006; Zhao, 2006; Stauffer, 1994; Denny, 1976; Dunaway, 1995; Cunniff, 1994; Moore, 1999; Agnarsson, 2010). A recent study has discovered a dragline silk that is twice as tough as any other previously described silk. This silk belongs to *Caerostris darwini* Kuntner & Agnarsson, a spider which constructs its orb web suspended above streams, rivers, and lakes (Agnarsson, 2010). To thoroughly understand all the various properties of spider silk, we must characterize the different kinds of silk.

The stress-strain behavior of the egg sac silk of *Araneus diadematus* (Van Nimmen, 2005b) presents a logarithmic behavior, which is completely different from the behavior of dragline and viscid silk. The same can be said about the egg sac silk of *Argiope bruennichi* (Zhao, 2006). The stress-strain curves of the egg sac silk start with a small elastic region and then present an extremely flat plastic-hardening region (Van Nimmen, 2005b). The breaking strain is roughly the same as that of the dragline, but tensile strength is about 3 to 4 times lower. Egg case silk has an initial modulus, a measurement of the stiffness of the fiber, which is significantly higher than that of dragline thread. These differences are partly due to the different amino acid compositions in the silks. To our knowledge, few studies have been conducted on stalks of egg sac silk. In general, each egg sac consists of two major parts that can be distinguished by the naked eye, an egg sac case and a stalk. The egg sac case houses eggs, while the stalk attaches the cocoon to the substrate (Hajer, 2009). In the literature, the strain of spider egg sac silk is in the range from 19 % for *Araneus gemmoides* (Stauffer, 1994) to 29 % for *Argiope argentata* (Blackledge, 2006), showing an average value of ~26 %; the average stress is 1.1 GPa with a minimum value of 0.3 GPa for *Araneus diadematus* (Van Nimmen, 2005b) and the maximum stress of 2.3 GPa for *Araneus gemmoides* (Stauffer, 1994).

One study pulled bundles of 100 dragline and minor ampullate silk threads at constant speeds (Stauffer, 1994). They observed that physical interactions between fibers influence elongation and thus increase the stretching capabilities of the bundle compared to that of a single fiber. They observe that *Nephila clavipes* dragline silk shows almost double the final stress value compared to the same silk of *Araneus gemmoides*, whereas the minor ampullate silk shows roughly the same final stress value (Stauffer, 1994).

The cave spider *Meta menardi* (Latreille) is generally found in dark and humid places like caves and mines throughout the northern hemisphere, from northern Europe to Korea and northern Africa (Platnick, 2011; Lepore, 2012b). *Meta menardi* has recently been chosen as the "European spider of the year 2012" from the European Society of Arachnology. Since no engineering studies of the egg sac of the cave spider *Meta menardi* yet exist and just a few studies have been focus on egg sacs, we conducted tensile tests on stalks of egg sac silk. We tested the stalk which connect the egg sacs of *Meta menardi* to the ceiling of caves (the arrow, in Fig.8.2, indicates such sample). In total 15, stalks were found and were pulled until they broke. Samples were viewed under FESEM to analyze the fracture surfaces and measure the diameter of the stalk. To visualize the thread stacking in each stalk, a Focused Ion Beam (FIB) was used to cut the stalk. Using FESEM micrographies of cross-sections of the FIB-cut stalk and the processing software ImageJ 1.41o, the real diameter and the exact number of single threads in each stalk were measured, improving previous accuracy (Poza, 2002). Thus, the stress-strain curves and the Weibull shape and scale parameters of the egg sac silk stalk of *Meta menardi* are determined here.

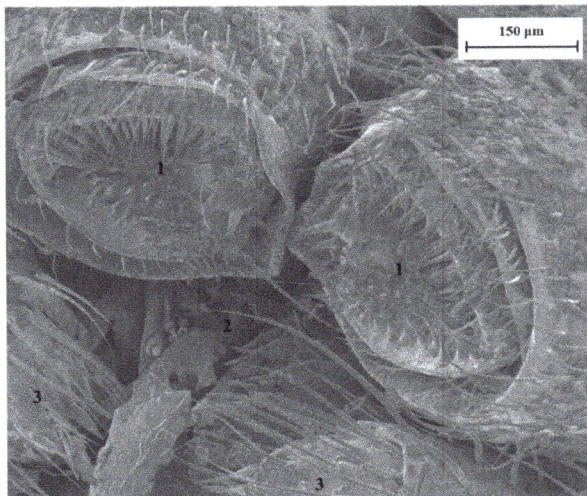

Figure 8.1 FESEM image of the spinnerets of *Meta menardi* (1. Anterior lateral; 2. Posterior median; 3. Posterior lateral).

Figure 8.2 Egg sac of the spider *Meta menardi*. Photo by Francesco Tomasinelli (2009).

8.2. Materials and Methods

Note that: no specific permits were required for the described field studies; the location is not privately-owned; the field studies did not involve endangered or protected species.

8.2.1. Tensile Testing

We identified caves in Piedmont (a north-western region of Italy) to search for *Meta menardi* egg sacs. The egg sacs are generally spun at the end of summer and hatch in late winter. Fifteen stalks of the egg sacs were taken from the caves in which they were found (Table 8.1). Since the egg sacs were collected in their natural habitat, the measured mechanical stress-strain behavior of the silk would most likely represent real silk characteristics more accurately than that produced by lab-reared spiders.

We collected fifteen stalks of the egg sacs in three different caves: four in Grotta Inferiore del Pugnetto, three in Grotta del Bandito, and eight in Grotta di

Chiabrano. The spiders of this species are generally found in dark areas close to cave openings, where temperature and humidity are still influenced by external conditions. The egg sacs hung from the ceilings of the caves with a bundle of threads (stalk) and were generally found in ventilated areas. The surveys were done on three separate days. The stalks of the egg sacs were carefully took from the cave ceilings and glued only by the ends to 30 mm x 50 mm cardboard holders. Each holder had a ~20 mm x 20 mm hole in its center so that the stalks could be suspended, transported while maintaining original tension, and mounted on the testing machine without being damaged. All tests were done in the Laboratory of Bio-inspired Nanomechanics "Giuseppe Maria Pugno" (Politecnico di Torino, Italy) with an air temperature of 22 ± 1 °C and 31 ± 2 % relative humidity.

Tensile tests were conducted on thirteen of the fifteen specimens, whereas the remaining two specimens were considered representative of the tested samples and examined under the FESEM and FIB. The tensile tests were conducted using a testing machine (Insight 1 kN, MTS, Minnesota, USA), equipped with a 10 N cell load with pneumatic clamps (closure pressure of 275.6 kPa). The cardboard holders were placed between the clamps with double-sided tape defining an initial length l_0 in the range from 18 to 19 mm. Once the holders were in place, the clamps were brought to zero tension and then the sides of the holders were cut, leaving the stalk loose between the clamps. The specimens were pulled until they completely broke at a constant rate of 2 mm/min, consistent with the parameter setting of previous studies (Giunea, 2003; Madsen, 1999; Van Nimmen, 2005b; Vollrath, 2001; Gosline, 1986; Zhao, 2006; Stauffer, 1994; Álvarez-Padilla, 2009).

Table 8.1 List of the caves visited for the collection of the samples with collection date and number of samples.

Cave name	Speleological cadastre number	Municipality	Province	Date	Number of samples
Grotta del Bandito	1002 Pi/CN	Roaschia	Cuneo	02/2011	3
Grotta inferiore del Pugnetto or Tana del lupo	1502 Pi/TO	Mezzenile	Torino	02/2011	4
Grotta di Chiabrano or Tuna del Diau	1621 Pi/TO	Perrero	Torino	02/2011	8

The computer program TestWorks 4 (MTS, Minnesota, USA) recorded the applied tensile force data, and the stress-strain curves were computed using the estimation of the real diameter and the exact number n of single threads at the cross-section of each stalk. Stress σ, strain ε, and modulus E were calculated using the following equations (1, 2, 3):

$$\sigma = \frac{F}{\sum\limits_{i=1}^{n} A_i} = \frac{F}{A_b} \tag{1}$$

$$\varepsilon = \frac{\Delta l}{l_0} \tag{2}$$

$$E = \frac{d\sigma}{d\varepsilon}\bigg|_0 \tag{3}$$

where F is the force measured by the testing machine, A_b is the initial cross-sectional area of the stalk (given as the initial cross-sectional area A_i of a single thread multiplied by the number n of stalk threads), l_0 is the initial length of the stalk, and Δl is the change in stalk length during the test. The area under the stress-strain curve gives the energy required to break the material, and this variable can be used to quantify toughness. Spider silk dissipates energy in volume, and so the classical fracture toughness cannot be defined, which suggest significant intrinsic toughening mechanisms.

Weibull statistics were applied to the stress results of the tensile tests. The probability of failure P for a stalk is defined as:

$$P(\sigma) = 1 - e^{-\left(\frac{\sigma}{\sigma_0}\right)^m} \tag{4}$$

where σ is the applied stress, m is the Weibull shape parameter, or Weibull modulus, and σ_0 is the Weibull scale parameter. The cumulative probability $P_i(\sigma_i)$ can be obtained experimentally as:

$$P_i(\sigma_i) = \frac{i - 1/2}{N} \tag{5}$$

where N is the total number of measured fracture stresses σ_i ranked in ascending order.

8.2.2. FESEM and FIB Characterization

Each stalk was cut by FIB (FEI Quanta 3D FEG, at 5 kV). The real diameter and the exact number of single threads in each stalk was determined using FESEM (FEI-Inspect™ F50, at 1 - 2 kV) micrographies of the FIB-cut stalk cross-section and the processing software ImageJ 1.41o.

8.3. Results

We performed tensile tests on the egg sac silk stalks of *Meta menardi*. The 13 stalks that we were tested were divided into two groups depending on the stalk type. Two types of stalk could be macroscopically distinguished: "cable" type (group A) and "ropey" type (group B). The "cable" type stalk consists of a series of threads tightly packed together to form a very compact structure (Fig.8.3 a), whereas the threads are not compact in the "ropey" type stalk (Fig.8.3 b). There were 4 stalks in group A and 6 stalks in group B. The remaining stalks were discarded as they did not reveal concrete information in terms of tensile strength. The performed tensile tests show significantly different values for stress, strain, and modulus; Weibull statistics were applied to interpret the results.

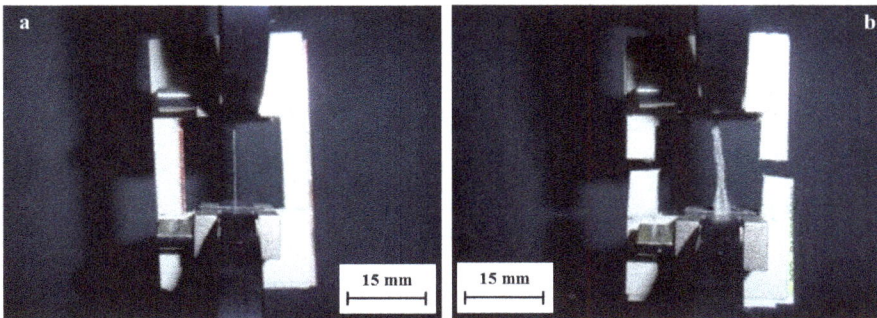

Figure 8.3 Distinction of the stalk types: cable-like (Group A) (a) and ropey (Group B) (b).

The FESEM images show that the threads composing the stalks are all of similar diameter and are parallel-oriented (Figs.8.4 a, b, c). Stalk ends were clamped between the pneumatic clamps with double-sided tape at a closure pressure of 275.6 kPa, which is a high pressure compared with the testing forces. Consequently, the macroscopic unraveling or slipping of the stalk or the cardboard holders between the clamps were impossible, so only sufficient bundle stretching is allowed. Moreover, no additional length is available for sliding after the clamps so we can exclude artifacts in our observations.

The diameter of a single thread is 6.03 µm (Fig.8.4 d), which is close to the upper value of the range from 1 to 6 µm previously indicated (Foelix, 1996), whereas the diameter of the bundles fall in the range from 200 to 300 µm. The FIB images allow us to observe how many silk threads compose each stalk and thus calculate the real thread cross-sectional area. Using the FESEM, we see that each stalk is composed of an average of 150 single silk threads, corresponding to an effective cross section of 4283.67 µm².

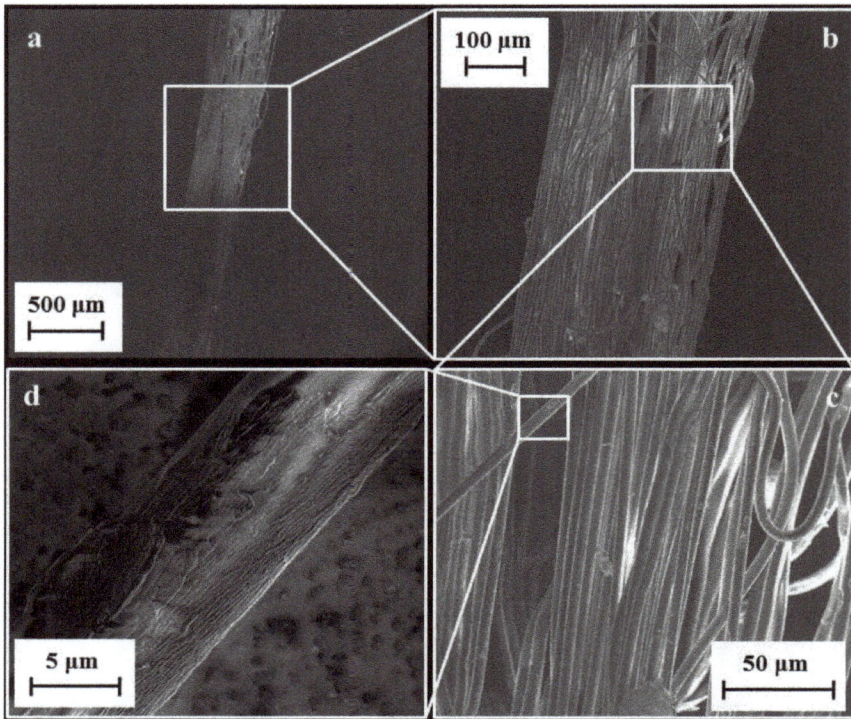

Figure 8.4 FESEM characterization of the silk stalk at different magnifications.

From the various tensile tests, and although the data is scattered, we calculate the average failure stress of 0.355 GPa for group A and 0.286 GPa for group B. The average failure strain is calculated to be 318 % for group A and 227 % for group B. The average values of toughness are 76.5 MJ/m³ for group A and 51.3 MJ/m³ for group B. The Young's modulus is calculated as the initial slope of the stress-strain curve, which is equal to 20.4 GPa for group A and 22.46 GPa for group B. Fig.8.5 a, b shows the various characterized stress-strain curves.

The stress-strain curves show different shapes, which is caused by the varying number of threads that compose each stalk. The curves show a small initial elastic region which reaches a maximum stress and then drops quickly to very low values. It continues to large strains until failure is reached, in some cases through a series of peaks caused by the breaking of single or a small number of threads in the stalk. The strain values differed but were all above 20 %, with some stalks reaching 300 % strain or more before breaking. Two tests were pulled to an extraordinary length; the maximum strain to which they were subjected was 751 % for stalk A2, corresponding to a toughness value of 130.7 MJ/m³ (solid line in Fig.8.5 a), and 721 % for stalk B4, corresponding to a toughness value of 117.4 MJ/m³ (solid line in Fig.8.5 b).

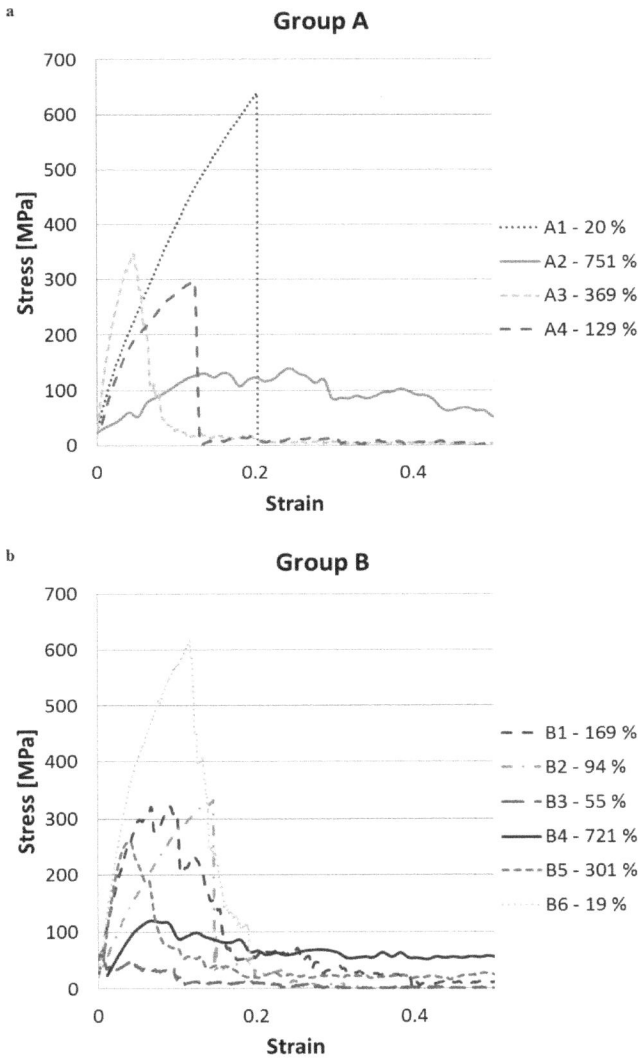

Figure 8.5 Stress-strain curves for group A (a) or B (b) stalks.

Following Weibull statistics, we apply Eq. (4) to the set of fracture stresses of the egg sac silk stalks of *Meta menardi* reported in Table 8.2. The Weibull modulus m, an index of the dispersion of the stress distribution, is 1.8 for group A (Fig.8.6 a) and 1.5 for group B (Fig.8.6 b), whereas σ_0, an index of the mean value of the stress distribution, is equal to 0.409 GPa for group A and 0.326 GPa for group B. Note that the correlation coefficient is high ($R^2 = 0.97$) for both the groups.

a

Group A

$y = 1.8423x + 1.6467$
$R^2 = 0.9721$

$\ln(-\ln(1-P_i))$

$\ln(\sigma_i)$

b

Group B

$y = 1.4719x + 1.6488$
$R^2 = 0.9659$

$\ln(-\ln(1-P_i))$

$\ln(\sigma_i))$

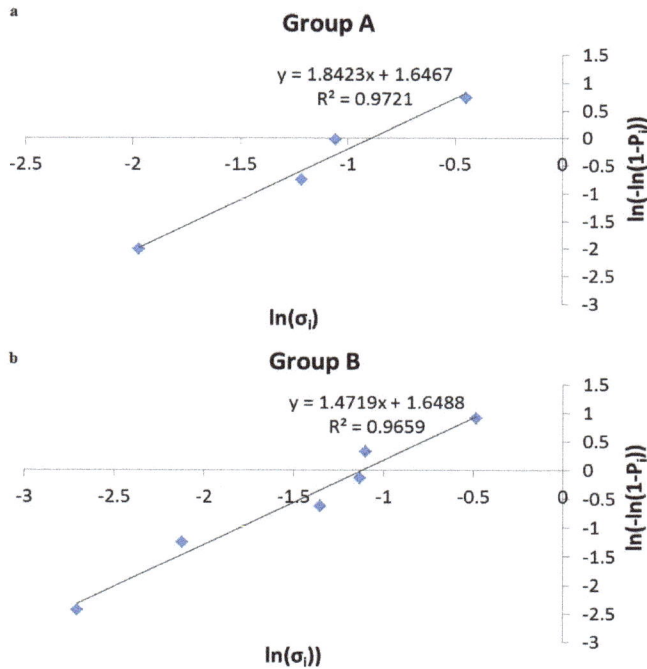

Figure 8.6 Weibull statistics for stress of group A (a) and B (b) stalks.

Table 8.2 The measured stress of each stalk in ascending order.

Group A		Group B	
Test n°	Stress (GPa)	Test n°	Stress (GPa)
1 (A2)	0.139	1 (B3)	0.067
2 (A3)	0.297	2 (B4)	0.120
3 (A4)	0.347	3 (B5)	0.259
4 (A1)	0.639	4 (B1)	0.322
		5 (B2)	0.332
		6 (B6)	0.617

8.4. Discussion

Previously, scientists have focused on different types of silk and mechanically characterized them. Only a few studies on tensile tests conducted on egg sac silk have been performed, particularly on *Argiope argentata* (Blackledge, 2006), *Araneus diadematus* (Van Nimmen, 2005b), *Nephila madagascariensis* (Gosline,

1986), *Argiope bruennichi* (Zhao, 2006), *Araneus gemmoides*, and *Nephila clavipes* (Stauffer, 1994). In addition, the genuses Nephila, Araneus, and Meta belong to three related families of orb web weavers (Nephilidae, Araneidae, and Tetragnathidae, respectively (Platnick, 2011)) and thus general trends can be seen (Xiao, 2008). The shapes of the observed stress-strain curves have a similar shape to that of carbon nanotube (CNT) bundles (Bosia, 2010; Perez-Rigueiro, 2000). These curves present a series of kinks or load drops, which are an indication of sub-bundle failures when a bundle is pulled parallel to its axis. Our data also have a series of kinks, indicating that the failure of the bundle at its peak load occurs with the fracture of sub-bundles. Although the curves are similar to those of CNT bundles, they are completely different from those of dragline silk bundles and egg sac silk stalks (Stauffer, 1994). Comparing their results to ours, we see that their failure stresses and toughness are much higher.

The β-sheet nanocrystals are held together by hydrogen bonds, one of the weakest chemical bonds. When a thread is pulled, the force peaks in the force-displacement graph confirm that the hydrogen bonds break and reform at an adjacent hydrogen bond ring. This occurs by preserving the initial side-chain orientation and shifting or by rotating and forming an opposite side-chain orientation. This bond breaking leads to a series of force peaks in the mechanical response and increases the total dissipated energy (Keten, 2010b). The size of the β-sheet nanocrystals influences the tensile response of a silk thread; consequently, the smaller the crystals, the greater the strength and toughness of the thread. As mentioned above, the fibers are made up of semi-amorphous α-chains and β-pleated sheets which are embedded in a rubber-like matrix. Images from the FESEM further show that the fibers consist of 2 layers (Vollrath, 1996): an inner layer and an outer coating. It seems that some fibers have a polymeric-like fracture surface and some have a more regular surface. This second case is probably due to the different crystals composing the fiber. In fact, β-sheets are crystal-like and are responsible for the toughness of the thread. They also have a more fragile rupture. On the other hand, we can assume that some fibers have a very ductile break caused by amorphous rubber-like regions (Fig.8.7 a, b).

Having cut our stalk with FIB, we are able to observe the cross section of the stalks at a SEM eye angle of 52° (Fig.8.8 a, b, c) and from the top (Fig.8.8 d). Each stalk is composed of a series of single silk threads which, when pulled, stack together to form what we initially hypothesized as a cylindrical cable. The diameters of the egg sac silk threads (~6 µm) are slightly smaller than those of egg sac silk of *Nephila clavipes* (~7 µm) (Stauffer, 1994) but equal to those of *Argiope bruennichi* (Zhao, 2006) and still much larger than the dragline silk (~1.4 µm) of the same species. For comparison, the diameters of dragline silk and minor ampullate in *Nephila clavipes* and *Araneus gemmoides* are estimated to be 3 and 2.5 µm (Stauffer, 1994), and 2.5 and 2 µm (Zhao, 2006), respectively.

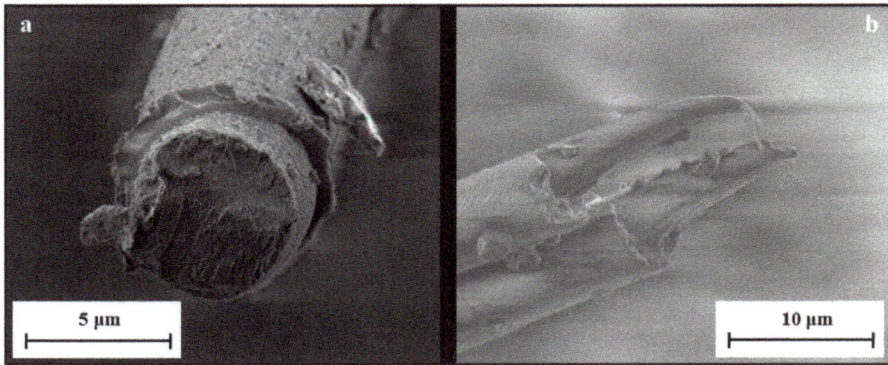

Figure 8.7 Detailed views of fracture surfaces of broken silk fibers.

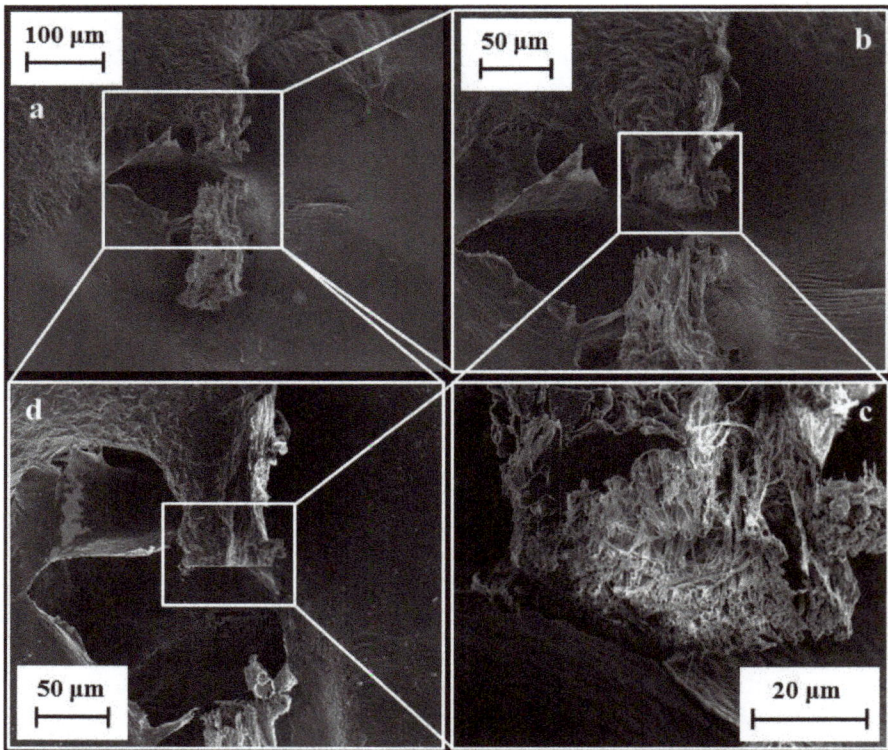

Figure 8.8 FESEM characterization of the stalk cut with FIB: (a, b, c) at an eye angle of 52° and (d) from the top.

The strains sustained by our fibers are impressively high; some stalks were pulled to more than 200 % their original length, reaching values of 721 - 751 %, which have not yet been observed in any spider single thread or stalk of egg sac silk. Such enormous elongations suggest a huge unrolling mechanism in the stalk.

In Figs.8.9-8.11, we report toughness, ultimate stress, and ultimate strain, respectively, for different types of spider silks; specifically, in Fig.8.11 our record of ultimate strain emerges. The reason for this very high strain is unknown but might be caused by an interaction and different disposition of the α-chains and β-pleated sheets within the fibers, thus allowing them to stretch to such high strain values. As stated in the introduction, it has been observed that physical interactions between fibers can influence the elongation data and increase the stretching capabilities of the stalk compared to those of a single fiber (Stauffer, 1994). We see that extreme strain of the stalks can be caused by macroscopic unraveling of the stalk itself. The failure strains of the egg sac silk of *Araneus diadematus* reached values of 30-40 %, much lower than the strains measured here (Van Nimmen, 2005b). Egg sac threads from *Nephila clavipes* extended to 24 ± 2 % their initial lengths and the maximum stress was observed to be 1.3 ± 0.2 GPa, whereas for *Araneus gemmoides*, these values were respectively 19 ± 2 % and 2.3 ± 0.2 GPa (Stauffer, 1994).

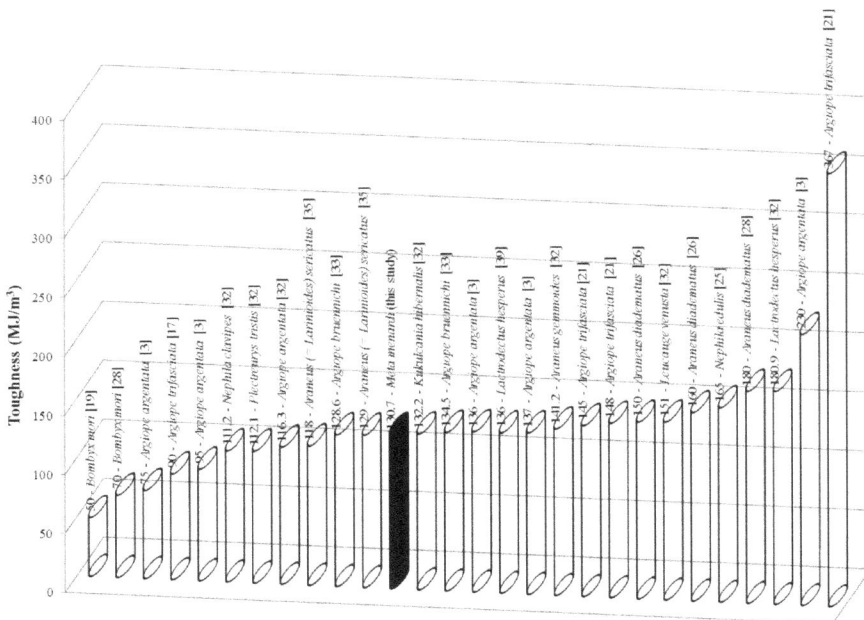

Figure 8.9 The maximum toughness of different types of (mainly spider) silks.

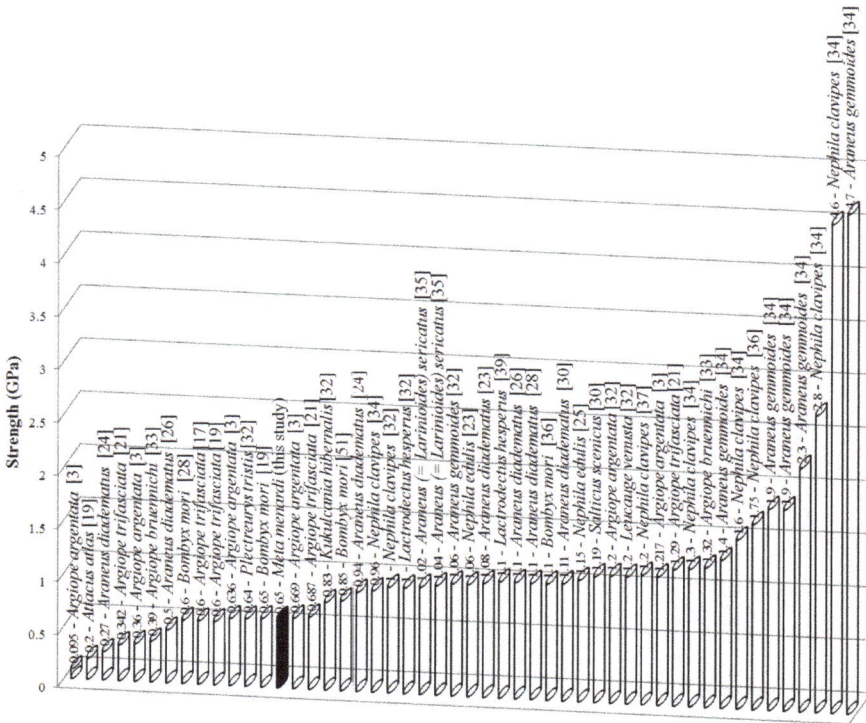

Figure 8.10 The maximum strength of different types of (mainly spider) silks.

The failure stresses of our stalks are much inferior to these, but the strains sustained by our stalks are much higher, which is most likely due to physical interactions within the stalks and the type of deformation occurring at the nanoscale. Bundles of dragline and minor ampullate silk composed of 100 threads were also tested (Stauffer, 1994) and show a wide range over which breaking occurred; these authors were not able to determine a useful value for the tensile strength of the fiber bundles due to the great variety in thread diameters in the bundle (Stauffer, 1994). We propose a solution to this problem using Weibull statistics, and our results (m is in the range from 1.5 to 1.8, and σ_0 is in the range from 0.33 to 0.41 GPa) are consistent with the values of the shape (m) and scale (σ_0) parameters of the Weibull parameters of 3.4 and 0.6 GPa for the dragline of *Argiope trifasciata* (Perez-Rigueiro, 2001; Perez-Rigueiro, 1998) and 5.7 and 0.4 GPa for the silkworm cocoons of *Bombyx mori* (Perez-Rigueiro, 1998), respectively.

The large standard deviation in the values of our stress and strain results within the two groups of stalks might be due to differences in terms of diameter,

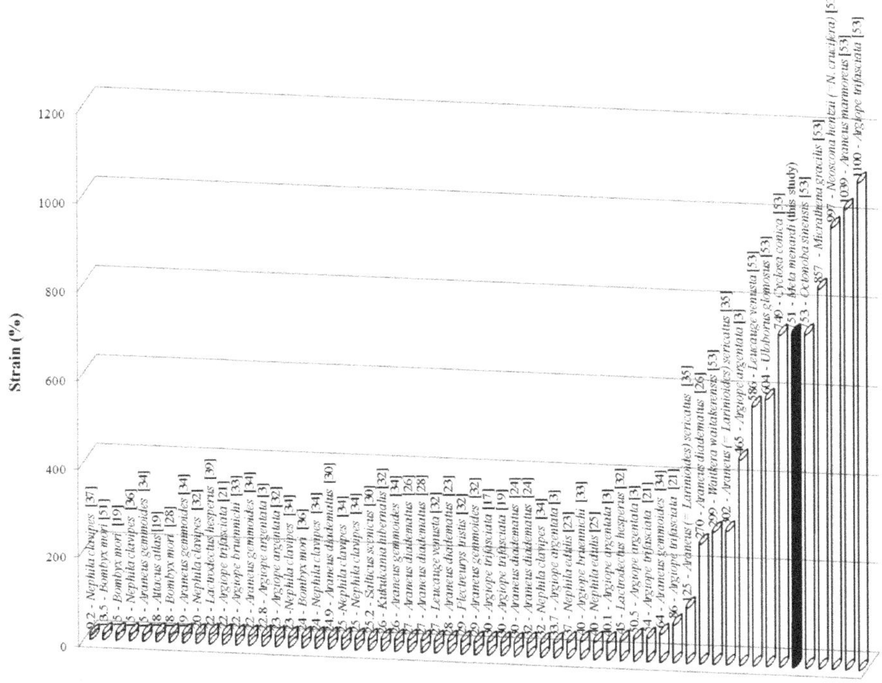

Figure 8.11 The maximum strain of different types of (mainly spider) silks, showing the record for egg sac silk stalks observed in the present study

number of threads, and the physical condition of the stalks, which may have affected performance (see Table 8.3). All stalks were obtained from the natural habitat of the spider where humidity and temperature play an important role. Moisture has been observed to induce thread supercontraction causing them to tighten up (Guinea, 2003). The temperature of the caves was roughly 9 ± 2°C, whereas the tests were performed in an environment at a much higher temperature, which might have caused the fibers to change from their natural state. The tests were also performed a couple of days after collection and were maintained in the laboratory at different conditions, which might have caused the modification or loss of some thread properties.

We observe that, the higher the stress sustained by the stalk, the lower the maximum strain before breakage. If strain reached high values the peak stresses did not exceed 0.64 GPa. In this case, we assume that the thread deformed in a rubber-like way, extending to significant lengths, due to physical interactions (Pugno, 2010; Allmeling, 2006; Dal Pra, 2005; Dal Pra, 2006) between the threads composing the stalk.

Table 8.3 The main parameters which may influence tensile testing results: systematics, function, silk-producing glands, temperature and humidity, initial length (l_0) of samples, number of tested threads, selected strain rate, and number of tested samples. Spider nomenclature according to (Platnick, 2011).

Ref.	Class, Order	Family	Species	Function	Glands	Experimental conditions (temperature and humidity)	l_0	Number of threads	Strain rate	Number of samples
[24]	Van Nimmen, 2005b	Arachnida, Araneae	Araneidae	Araneus diadematus	egg sac	tubuliform (cylindrical)	wet (20 °C, 65 %)	20 mm	1	20 mm/min	398
[24]	Van Nimmen, 2005b	Arachnida, Araneae	Araneidae	Araneus diadematus	structural threads and dragline	major ampullate	wet (20 °C, 65 %)	20 mm	1	20 mm/min	183
[26]	Gosline, 1999	Arachnida, Araneae	Araneidae	Araneus diadematus	structural threads and dragline	major ampullate	not given	not given	not given	not given	not given
[28]	Römer, 2008	Arachnida, Araneae	Araneidae	Araneus diadematus	structural threads and dragline	major ampullate	not given	not given	not given	not given	not given
[23]	Madsen, 1999	Arachnida, Araneae	Araneidae	Araneus diadematus	structural threads and dragline	major ampullate	wet (24 °C, 50 %)	6.9 mm	1	3 mm/min	30
[30]	Ortlepp, 2008	Arachnida, Araneae	Araneidae	Araneus diadematus	structural threads and dragline	major ampullate	not given	not given	2	14-20 mm/min	16
[26]	Gosline, 1999	Arachnida, Araneae	Araneidae	Araneus diadematus	glue coating on viscid capture threads	aggregate glands	not given	not given	not given	not given	not given
[34]	Stauffer, 1994	Arachnida, Araneae	Araneidae	Araneus gemmoides	auxiliary radial threads and temporary scaffolding	minor ampullate	not given	40 mm	1	5 mm/min	10
[34]	Stauffer, 1994	Arachnida, Araneae	Araneidae	Araneus gemmoides	egg sac	tubuliform (cylindrical)	not given	40 mm	1	5 mm/min	10
[34]	Stauffer, 1994	Arachnida, Araneae	Araneidae	Araneus gemmoides	structural threads and dragline	major ampullate	not given	40 mm	1	5 mm/min	10

continued **Table 8.3** The main parameters which may influence tensile testing results: systematics, function, silk-producing glands, temperature and humidity, initial length (l_0) of samples, number of tested threads, selected strain rate, and number of tested samples. Spider nomenclature according to (Platnick, 2011).

Ref.	Class, Order	Family	Species	Function	Glands	Experimental conditions (temperature and humidity)	l_0	Number of threads	Strain rate	Number of samples
[32] Swanson, 2006	Arachnida, Araneae	Araneidae	Araneus gemmoides	structural threads and dragline	major ampullate	not given	not given	1	1 % strain/s	23
[34] Stauffer, 1994	Arachnida, Araneae	Araneidae	Araneus gemmoides	auxiliary radial threads and temporary scaffolding	minor ampullate	not given	40 mm	100	5 mm/min	10
[34] Stauffer, 1994	Arachnida, Araneae	Araneidae	Araneus gemmoides	structural threads and dragline	major ampullate	not given	40 mm	100	5 mm/min	10
[35] Denny, 1976	Arachnida, Araneae	Araneidae	Araneus (= Larinioides) sericatus	structural threads and dragline	major ampullate	wet (21 °C, 50 %)	25 mm	1	13.2 mm/min	60
[35] Denny, 1976	Arachnida, Araneae	Araneidae	Araneus (= Larinioides) sericatus	glue coating on viscid capture threads	aggregate glands	wet (21 °C, 50 %)	25 mm	1	217.1 mm/min	41
[53] Opell, 2000	Arachnida, Araneae	Araneidae	Araneus marmoreus	adhesive threads of the catching spiral	flagelliform glands	wet (23 °C, 60 %)	20 mm	1	0.017 mm/min	3
[3] Blackledge, 2006	Arachnida, Araneae	Araneidae	Argiope argentata	structural threads and dragline	major ampullate	wet (21.5 °C, 45 %)	21 mm	1	12.6 mm/min (1 % strain/s)	13
[32] Swanson, 2006	Arachnida, Araneae	Araneidae	Argiope argentata	structural threads and dragline	major ampullate	not given	not given	1	1 % strain/s	62
[3] Blackledge, 2006	Arachnida, Araneae	Araneidae	Argiope argentata	wrapping silk and packing silk	aciniform gland	wet (21.5 °C, 45 %)	10 mm	1	6 mm/min (1 % strain/s)	28

continued **Table 8.3** The main parameters which may influence tensile testing results: systematics, function, silk-producing glands, temperature and humidity, initial length (l_0) of samples, number of tested threads, selected strain rate, and number of tested samples. Spider nomenclature according to (Platnick, 2011).

Ref.	Class, Order	Family	Species	Function	Glands	Experimental conditions (temperature and humidity)	l_0	Number of threads	Strain rate	Number of samples
[3]	Arachnida, Araneae	Araneidae	Argiope argentata	adhesive capture threads of the catching spiral	flagelliform glands	wet (21.5 °C, 45 %)	21 mm	1	12.6 mm/min (1 % strain/s)	87
[3]	Arachnida, Araneae	Araneidae	Argiope argentata	auxiliary radial threads and temporary scaffolding	minor ampullate	wet (21.5 °C, 45 %)	21 mm	1	12.6 mm/min (1 % strain/s)	51
[3]	Arachnida, Araneae	Araneidae	Argiope argentata	egg sac	tubuliform (cylindrical)	wet (21.5 °C, 45 %)	21 mm	1	12.6 mm/min (1 % strain/s)	29
[33]	Arachnida, Araneae	Araneidae	Argiope bruennichi	egg sac	tubuliform (cylindrical)	wet (24 °C, 34 %)	20 mm	1	10 mm/min	4
[33]	Arachnida, Araneae	Araneidae	Argiope bruennichi	structural threads and dragline	major ampullate	wet (24 °C, 34 %)	20 mm	1	10 mm/min	4
[17]	Arachnida, Araneae	Araneidae	Argiope trifasciata	structural threads and dragline	major ampullate	wet (20 °C, 60 %)	20 mm	1	0.24 mm/min (2*10⁻⁴ /s)	28
[21]	Arachnida, Araneae	Araneidae	Argiope trifasciata	structural threads and dragline	major ampullate	not given	21 mm	1	12.6 mm/min (1 % strain/s)	7
[19]	Arachnida, Araneae	Araneidae	Argiope trifasciata	structural threads and dragline	major ampullate	wet (20 °C, 60 %)	10 mm	1	0.12 mm/min (2*10⁻⁴ /s)	10

continued **Table 8.3** The main parameters which may influence tensile testing results: systematics, function, silk-producing glands, temperature and humidity, initial length (l_0) of samples, number of tested threads, selected strain rate, and number of tested samples. Spider nomenclature according to (Platnick, 2011).

Ref.	Class, Order	Family	Species	Function	Glands	Experimental conditions (temperature and humidity)	l_0	Number of threads	Strain rate	Number of samples
[21]	Arachnida, Araneae	Araneidae	*Argiope trifasciata*	wrapping silk and packing silk	aciniform gland	not given	10 mm	2	6 mm/min (1 % strain/s)	14
[21]	Arachnida, Araneae	Araneidae	*Argiope trifasciata*	auxiliary radial threads and temporary scaffolding	minor ampullate	not given	21 mm	2	12.6 mm/min (1 % strain/s)	11
[53]	Arachnida, Araneae	Araneidae	*Argiope trifasciata*	adhesive threads of the catching spiral	flagelliform glands	wet (23 °C, 60 %)	20 mm	1	0.017 mm/min	3
[53]	Arachnida, Araneae	Araneidae	*Micrathena gracilis*	adhesive threads of the catching spiral	flagelliform glands	wet (24 °C, 60 %)	20 mm	1	0.017 mm/min	3
[53]	Arachnida, Araneae	Araneidae	*Neoscona hentzii* (= *N. crucifera*)	adhesive threads of the catching spiral	flagelliform glands	wet (23 °C, 60 %)	20 mm	1	0.017 mm/min	3
[53]	Arachnida, Araneae	Araneidae	*Cyclosa conica*	adhesive threads of the catching spiral	flagelliform glands	wet (25 °C, 61 %)	20 mm	1	0.017 mm/min	3
[53]	Arachnida, Araneae	Uloboridae	*Octonoba sinensis*	dry cribellar capture threads of the catching spiral	flagelliform glands	wet (23 °C, 61 %)	20 mm	1	0.017 mm/min	3
[53]	Arachnida, Araneae	Uloboridae	*Uloborus glomosus*	dry cribellar capture threads of the catching spiral	flagelliform glands	wet (24 °C, 62 %)	20 mm	1	0.017 mm/min	3

continued **Table 8.3** The main parameters which may influence tensile testing results: systematics, function, silk-producing glands, temperature and humidity, initial length (l_0) of samples, number of tested threads, selected strain rate, and number of tested samples. Spider nomenclature according to (Platnick, 2011).

Ref.	Class, Order	Family	Species	Function	Glands	Experimental conditions (temperature and humidity)	l_0	Number of threads	Strain rate	Number of samples
[53] Opell, 2000	Arachnida, Araneae	Uloboridae	Waitkera waitakerensis	dry cribellar capture threads of the catching spiral	flagelliform glands	wet (25 °C, 70 %)	20 mm	1	0.017 mm/min	3
[32] Swanson, 2006	Arachnida, Araneae	Filistatidae	Kukulcania hibernalis	dragline	major ampullate	not given	not given	1	1 % strain/s	102
[32] Swanson, 2006	Arachnida, Araneae	Theridiidae	Lactrodectus hesperus	structural threads and dragline	major ampullate	not given	not given	1	1 % strain/s	70
[39] Moore, 1999	Arachnida, Araneae	Theridiidae	Lactrodectus hesperus	auxiliary radial threads and temporary scaffolding	minor ampullate	not given	12 mm	30	12.6 mm/min	30
[32] Swanson, 2006	Arachnida, Araneae	Tetragnathidae	Leucauge venusta	structural threads and dragline	major ampullate	not given	not given	1	1 % strain/s	61
[53] Opell, 2000	Arachnida, Araneae	Tetragnathidae	Leucauge venusta	adhesive threads of the catching spiral	flagelliform glands	wet (25 °C, 60 %)	20 mm	1	0.017 mm/min	3
this study	Arachnida, Araneae	Tetragnathidae	Meta menardi	egg sac	tubuliform (cylindrical)	wet (22 °C, 31 %)	18-19 mm	150	2 mm/min	10
[34] Stauffer, 1994	Arachnida, Araneae	Nephilidae	Nephila clavipes	auxiliary radial threads and temporary scaffolding	minor ampullate	not given	40 mm	1	5 mm/min	10
[34] Stauffer, 1994	Arachnida, Araneae	Nephilidae	Nephila clavipes	egg sac	tubuliform (cylindrical)	not given	40 mm	1	5 mm/min	10

continued **Table 8.3** The main parameters which may influence tensile testing results: systematics, function, silk-producing glands, temperature and humidity, initial length (l_0) of samples, number of tested threads, selected strain rate, and number of tested samples. Spider nomenclature according to (Platnick, 2011).

Ref.	Class, Order	Family	Species	Function	Glands	Experimental conditions (temperature and humidity)	l_0	Number of threads	Strain rate	Number of samples
[34] Stauffer, 1994	Arachnida, Araneae	Nephilidae	Nephila clavipes	structural threads and dragline	major ampullate	not given	40 mm	1	5 mm/min	10
[32] Swanson, 2006	Arachnida, Araneae	Nephilidae	Nephila clavipes	structural threads and dragline	major ampullate	not given	not given	1	1 % strain/s	66
[36] Dunaway, 1995	Arachnida, Araneae	Nephilidae	Nephila clavipes	structural threads and dragline	major ampullate	wet (23 °C, 49 %)	12.7 mm	1	12.7 mm/min (100 %/min)	19
[37] Cunniff, 1994	Arachnida, Araneae	Nephilidae	Nephila clavipes	structural threads and dragline	major ampullate	wet (21 °C, 50 %)	50.8 mm	1	304.8 mm/min (10 % strain/s)	30
[34] Stauffer, 1994	Arachnida, Araneae	Nephilidae	Nephila clavipes	auxiliary radial threads and temporary scaffolding	minor ampullate	not given	40 mm	100	5 mm/min	10
[34] Stauffer, 1994	Arachnida, Araneae	Nephilidae	Nephila clavipes	structural threads and dragline	major ampullate	not given	40 mm	100	5 mm/min	10
[23] Madsen, 1999	Arachnida, Araneae	Nephilidae	Nephila edulis	structural threads and dragline	major ampullate	wet (24 °C, 50 %)	6.9 mm	1	3 mm/min	30
[25] Vollrath, 2001	Arachnida, Araneae	Nephilidae	Nephila edulis	structural threads and dragline	major ampullate	wet (22 °C, 50 %)	12 mm	1	6 mm/min (50 % strain/min)	not given
[32] Swanson, 2006	Arachnida, Araneae	Plectreuridae	Plectreurys tristis	dragline	major ampullate	not given	not given	1	1 % strain/s	108

Table 8.3 The main parameters which may influence tensile testing results: systematics, function, silk-producing glands, temperature and humidity, initial length (l_0) of samples, number of tested threads, selected strain rate, and number of tested samples. Spider nomenclature according to (Platnick, 2011).

continued

Ref.	Class, Order	Family	Species	Function	Glands	Experimental conditions (temperature and humidity)	l_0	Number of threads	Strain rate	Number of samples
[30]	Arachnida, Araneae	Salticidae	*Salticus scenicus*	dragline	major ampullate	not given	not given	2	14-20 mm/min	5
[19]	Insecta, Lepidoptera	Saturniidae	*Attacus atlas*	cocoon	silk glands	wet (20 °C, 60 %)	30 mm	1	0.36 mm/min (2*10⁻⁴ /s)	10
[28]	Insecta, Lepidoptera	Bombycidae	*Bombyx mori*	cocoon	silk glands	not given	not given	not given	not given	not given
[19]	Insecta, Lepidoptera	Bombycidae	*Bombyx mori*	cocoon	silk glands	wet (20 °C, 60 %)	30 mm	1	0.36 mm/min (2*10⁻⁴ /s)	10
[51]	Insecta, Lepidoptera	Bombycidae	*Bombyx mori*	cocoon	silk glands	wet (20 °C, 60 %)	40 mm	1	0.48 mm/min (0.0002 /s)	10
[36]	Insecta, Lepidoptera	Bombycidae	*Bombyx mori*	cocoon	silk glands	wet (23 °C, 49 %)	12.7 mm	1	3.81 mm/min (30 %/min)	20

8.5. Conclusion

The tensile properties and the Weibull shape and scale parameters of egg sac silk stalks of *Meta menardi*, obtained directly from their natural habitat, were determined in the present study. The results differ significantly compared with other tensile tests on spider silk. When compared with egg sac silk from other species of orbweb weavers, dragline silk, or minor ampullate silk, the presented results for maximum strain are much higher, up to 750 % than those previously reported, suggesting the discovery of the most stretchable egg sac silk stalk ever tested. Such significant elongations suggests a huge unrolling microscopic mechanism of the macroscopic stalk that, as a continuation of the protective egg sac, is expected to be composed by of very densely and randomly packed fibers.

Bibliography

Adamson, A. V. (1990). *Physical chemistry of surfaces.* Wiley: NY.

Agnarsson, I., Kuntner, M., & Blackledge, T. A. (2010). Bioprospecting finds the toughest biological material: extraordinary silk from a giant riverine orb spider. *PLoS ONE, 5,* 1-8.

Allmeling, C., Jokuszies, A., Reimers, K., et al. (2006). Use of spider silk fibres as an innovative material in a biocompatible artificial nerve conduit. *Journal of Cellular and Molecular Medicine,10,* 770-777.

Álvarez-Padilla, F., Dimitrov, D., Giribet, G., et al. (2009). Phylogenetic relationships of the spider family Tetragnathidae (Araneae, Araneoidea) based on morphological and DNA sequence data. *Cladistics, 25,*109-146.

Aristotle, *Historia Animalium,* 343 B.C. See Book IX, Part 9, translated by Thompson DAW, http://classics.mit.edu/Aristotle/history_anim.html.

Arnold, J. W. (1974). Adaptive features on the tarsi of cockroaches (Insecta: Dictyoptera). *International Journal of Insect Morphology and Embryology, 3,* 317-334.

Arzt, E., Gorb, S., & Spolenak, R. (2003). From micro to nano contacts in biological attachment devices. *Proceedings of the National Academy of Sciences, 100,* 10603-10606.

Schmidt, H. R. (1904). *Jena. Z. Naturwiss, 39,* 551-563.

Autumn, K., Liang, Y. A., Hsieh, S. T., et al. (2000). Adhesive force of a single gecko foot-hair. *Nature, 405,* 681-685.

Autumn, K. & Peattie, A. M. (2002a). Mechanism of adhesion in geckos. *Integrative and Comparative Biology, 42,* 1031-1090.

Autumn, K., Sitti, M., Liang, Y. A., et al. (2002b). *Proceedings of the National Academy of Sciences, 99,* 12252-12256.

Autumn, K., Ryan, M. J., & Wake, D. B. (2002c). Integrating historical and mechanistic biology enhances the study of adaptation. *The Quarterly Review of Biology, 77,* 383-408.

Autumn, K. (2006a). *Biological Adhesives* (ed. A. Smith and J. Callow), 225-255. Berlin Heidelberg: Springer Verlag.

Autumn, K., Dittmore, A., Santos, D., et al. (2006b). Frictional adhesion: a new angle on gecko attachment. *Journal of Experimental Biology, 209,* 3569-3579.

Autumn, K., Hsieh, S. T., Dudek, D. M., et al. (2006c). Dynamics of geckos running vertically. *Journal of Experimental Biology, 209,* 260-272.

Autumn, K., Majidi, C., Groff, R., et al. (2006e). Effective elastic modulus of isolated gecko setal arrays. *Journal of Experimental Biology, 209,* 3558-3568.

Autumn, K. (2007). Properties, principles and parameters of the gecko adhesive system, in Biological Adhesives, A. Smith and J. Callow (Eds.) (Springer Verlag, Berlin and Heidelberg, 2006), pp. 225–255; Autumn, K., "Gecko adhesion: structure, function and applications", *MRS Bulletin, 32,* 473-478.

Autumn, K. & Gravish, N. (2008). Gecko adhesion: evolutionary nanotechnology. *Philosophical Transactions of the Royal Society of London, Series A: Mathematical, Physical, and Engineering Sciences, 366,* 1575-1590.

Aksak, B., Murphy, M. P., & Sitti, M. (2007). Adhesion of biologically inspired vertical and angled polymer microfiber arrays. *Langmuir, 23,* 3322.

Barthlott, W. (1981). Epidermal and seed surface characters of plants: Systematic applicability and some evolutionary aspects. *Nordic Journal of Botany, 1,* 345-55.

Barthlott, W., & Neinhuis, C. (1997). Purity of the sacred lotus, or escape from contamination in biological surfaces. *Planta, 202,* 1-8.

Bell, W. J., Roth, L. M., & Nalepa, C. A. (2007). Cockroaches. Ecology behavior and natural history, in *The Johns Hopkins University Press,* Baltimore.

Bellucci, S., Bergamaschi, A., Bottini, M., et al. (2007). Biomedical platforms based on composite nanomaterials and cellular toxicity. *Journal of Physics, 61*, 95-98.

Bergmann, P. J., & Irschick, D. J. (2005). Effects of temperature on maximum clinging ability in a diurnal gecko: evidence for a passive clinging mechanism? *Journal of Experimental Zoology A, 303*, 785-791.

Beutel, R. G., & Gorb, S.N. (2001). Ultrastructure of attachment specializations of hexapods (Arthropoda): evolutionary patterns inferred from a revised ordinal phylogeny. *Journal of Zoological Systematics and Evolutionary Research, 39*, 177-207.

Brushan, B. (1999). *Principles and applications of tribology.* Wiley, NY.

Brushan, B. (2002). *Introduction to tribology.* Wiley, NY.

Bhushan, B., & Jung, Y. C. (2007a). Wetting study of patterned surfaces for superhydrophobicity. *Ultramicroscopy, 107*, 1033-41.

Bhushan, B., Nosonovsky, M., & Jung, Y. C. (2007b). Towards optimization of patterned superhydrophobic surfaces. *Journal of The Royal Society Interface, 4*, 643-8.

Bhushan, B., & Jung, Y. C. (2008). Wetting, adhesion and friction of superhydrophobic and hydrophilic leaves and fabricated micro/nanopatterned surfaces. *Journal of Physics: Condensed Matter, 20*, 225010 (24pp).

Bignardi, C., Petraroli, M., & Pugno, N. (2010). Nanoindentations on conch shells of Gastropoda and Bivalvia molluscs reveal anisotropic evolution against external attacks. *Journal of Nanoscience and Nanotechnology, 10*, 6453-6460.

Bitar, L. A., Voigt, D., Zebitz, C. P. W., et al. (2010). Attachment ability of the codling moth Cydia pomonella L. to rough substrates. *Journal of Insect Physiology, 56*, 1966-1972.

Blackledge, T. A., & Hayashi, C. Y. (2006). Silken toolkits: biomechanics of silk fibers spun by the orb web spider *Argiope argentata* (Fabricius 1775). *Journal of Experimental Biology, 209*, 2452-2461.

Blackledge, T. A., Kuntner, M., & Agnarsson, I. (2011). The Form and Function of Spider Orb Webs: Evolution from Silk to Ecosystems. Burlington: Academic Press. *Advances in Insect Physiology, 41*, 175-262.

Bosia, F., Pugno, N., & Buehler, M. (2010). Hierarchical simulations for the design of supertough nanofibres inspired by spider silk. *Physical Review E, 82,* 056103 (7pp).

Boutry, C., Řezáč, M., & Blackledge, T. A. (2011). Plasticity in Major Ampullate Silk Production in Relation to Spider Phylogeny and Ecology. *PLoS ONE, 6,* 1-8.

Brainerd, E. L. (1994). The evolution of lung-gill bimodal breathing and the homology of vertebrate respiratory pumps. *American Zoology, 34,* 128A.

Briggs, G. A. D., & Briscoe, B. J. (1977). The effect of surface topography on the adhesion of elastic solids. *Journal of Physics D: Applied Physics, 10,* 2453-2466.

Brunetta, L., & Craig, C. L. (2010). Spider Silk, Evolution and 400 Million Years of Spinning, Waiting, Snagging, and Mating. Csiro Publishing.

Buehler, M. J., Yao, H., Gao. H., et al. (2006). Cracking and adhesion at small scales: atomistic and continuum studies of flaw tolerant nanostructures. *Modelling and Simulation in Materials Science and Engineering, 14,* 799-816.

Bullock, J. M. R., Drechsler, P., & Federle, W. (2008). Comparison of smooth and hairy attachment pads in insects: friction, adhesion and mechanisms for direction-dependence. *Journal of Experimental Biology, 211,* 3333-3343.

Bullock, J. M. R., & Federle, W. (2009). Division of labour and sex differences between fibrillar, tarsal adhesive pads in beetles: effective elastic modulus and attachment performance. *Journal of Experimental Biology, 212,* 1876-1888.

Bullock, J. M. R., & Federle, W. (2010). The effect of surface roughness on claw and adhesive hair performance in the dock beetle *Gastrophysa viridula. Insect Science, 00,* 1-7.

Burton, Z., & Bhushan, B. (2005). Hydrophobicity, adhesion and friction properties of nanopatterened polymers and scale dependence for MEMS/NEMS. *Nano Letters, 5,* 1607-13.

Cartier, O. (1872). Studien über den feineren Bau der Haut bei den Reptilien. *Verhandl. Würz Phys-med Gesell, 1,* 83-96.

Cassie, A. B. D., & Baxter, S. (1944). Wettability of porous surfaces. *Transactions of the Faraday Society, 40,* 546-51.

Cheng, Y. T., & Rodak, D. E. (2005). Is the lotus leaf superhydrophobic? *Applied Physics Letters, 86*, 144101-3.

Cheng, Y. T., Rodak, D. E., Wong, C.A., et al. (2006). Effects of micro- and nano-structures on the self-cleaning behaviour of lotus leaves. *Nanotechnology, 17*, 1359-1362.

Clemente, C. J., & Federle, W. (2008). Pushing versus pulling: division of labour between tarsal attachment pads in cockroaches. *Proceedings of the royal society B, 275*, 1329-1336.

Clemente, C. J., Dirks, J. -H., Barbero, D. R., et al. (2009). Friction ridges in cockroach climbing pads: anisotropy of shear stress measured on transparent, microstructured substrates. *Journal of Comparative Physiology A, 195*, 805-814.

Coulson, S. R., Woodward, I., & Badyal, J. P. S. (2000). Superrepellent composite fluoropolymer surfaces. *Journal of Physical Chemistry B, 104*, 8836-40.

Craig, C. L. (2003). Spider Webs and Silk: Tracing Evolution from Molecules to Genes to Phenotypes. New York: Oxford University Press.

Cunniff, P. M., Fossey, S. A., Auerbach, M. A., et al. (1994). Mechanical and thermal properties of dragline silk from the spider *Nephila clavipes. Polymers for Advanced Technologies, 5*, 401-410.

Dai, Z., Gorb, S. N., & Schwarz, U. (2002). Roughness-dependent friction force of the tarsal claw systemin the beetle *Pachnoda marginata* (Coleoptera Scarabaeidae). *Journal of Experimental Biology, 205*, 2479-2485.

Dal Pra, I., Freddi, G., Minic, J., et al. (2005). *De novo* engineering of reticular connective tissue in vivo by silk fibroin nonwoven materials. *Biomaterials, 26*, 1987-1999.

Dal Pra, I., Chiarini, A., Boschi, A., et al. (2006). Novel dermoepidermal equivalents on silk fibroin-based formic acid-crosslinked threedimensional nonwoven devices with prospective applications in human tissue engineering/regeneration/repair. *International Journal of Molecular Medicine, 18*, 241-247.

Dellit, W. D. (1934). Zur Anatomie und Physiologie der Geckozehe. *Jena. Z. Naturwiss. 68*, 613-658.

Denny, M. (1976). The physical properties of spider's silk and their role in the design of orb-webs. *Journal of Experimental Biology, 65*, 483-506.

Dixon, A. F. G., Croghan, P. C., & Gowing, R. P. (1990). The mechanism by which aphids adhere to smooth surfaces. *Journal of Experimental Biology, 152*, 243-253.

Drechsler, P., & Federle, W. (2006). A multi-axis force sensor for studying insect biomechanics. *Journal of Comparative Physiology A, 192*, 1213-1222.

Dunaway, D. L., Thiel, B. L., & Viney, C. (1995). Tensile mechanical property evaluation of natural and epoxide-treated silk fibers. *Journal of Applied Polymer Science, 58*, 675-683.

Eberhard, W. G. (2010). Possible functional significance of spigot placement on the spinnerets of spiders. *Journal of Arachnology, 38*, 407-414.

Eigenbrode, S. D., & Jetter, R. (2002). Attachment to plant surface waxes by an insect predator. *Integrative and Comparative Biology, 42*, 1091-1099.

Eisner, T., & Aneshansley, D. J. (2000). Defense by foot adhesion in a beetle (*Hemisphaerota cyanea*). *Proceedings of the National Academy of Sciences, 97*, 6568-6573.

Elices, M., Pérez-Rigueiro, J., Plaza, G. R., et al. (2005). Finding inspiration in *Argiope trifasciata* spider silk fibers. *Journal of the Minerals Metals & Materials Society, 57*, 60-66.

Erbil, H. Y., Demirel, A. L., & Avci, Y. (2003). Transformation of a simple plastic into a superhydrophobic surface. *Science, 299*, 1377-80.

Federle, W., Rohrseitz, K., & Hölldobler, B. (2000). Attachment forces of ants measured with a centrifuge: better 'wax-runners' have a poorer attachment to a smooth surface. *Journal of Experimental Biology, 203*, 505-512.

Federle, W., Riehle, M., Curtis, A. S. G., et al. (2002). An integrative study of insect adhesion: Mechanics and wet adhesion of pretarsal pads in ants. *Integrative and Comparative Biology, 42*, 1100-1106.

Federle, W., Baumgartner W., & Hölldobler, B. (2003). Biomechanics of ant adhesive pads: frictional forces are rate- and temperature dependent. *Journal of Experimental Biology, 206*, 67-74.

Feng, L., Li, S., Li, Y., et al. (2002). Super-hydrophobic surfaces: From natural to artificial. *Advanced Materials, 14*, 1857-60.

Foelix, R. F. (1996). Biology of Spiders. New York: Oxford University Press.

Foradori, M. J., Kovoor, J., Moon, M. J., et al. (2002). Relation between the outer cover of the egg case of *Argiope aurantia* (Araneae: Araneidae) and the emergence of its Spiderlings. *Journal of Morphology, 252*, 218-226.

Fuller, K. N. G., & Tabor, D. (1975). The effect of surface roughness on the adhesion of elastic solids. *Proceedings of the Royal Society London, Series A, 345*, 327-342.

Furstner, R., Barthlott, W., Neinhuis, C., et al. (2005). Wetting and self-cleaning properties of artificial superhydrophobic surfaces. *Langmuir, 21*, 956-961.

Gadow, H. (1902). *Amphibia and Reptiles* (Macmillan & Company, Ltd., London).

Gao, H. J., Wang, X., Yao, H., et al. (2005). Mechanics of hierarchical adhesion structures of geckos. *Mechanics of Materials, 37*, 275-285.

Gennaro, J. G. J. (1969). The gecko grip. *Journal of Natural History, 78*, 36-43.

Gheysens, T., Beladjal, L., Gellynck, K., et al. (2005). Egg sac structure of *Zygiella x-notata* (Arachnida, Araneidae). *Journal of Arachnology, 33*, 549-557.

Gorb, S. N. & Scherge, M. (2000). Biological microtribology: anisotropy in frictional forces of orthopteran attachment pads reflects the ultrastructure of a highly deformable material. *Proceedings of the Royal Society London, Series B, 267*, 1239-1244.

Gorb, S. N. (2001a). Attachment Devices of Insect Cuticle. Dordrecht: Kluwer Academic Publishers.

Gorb, S., Gorb, E., & Kastner, V. (2001b). Scale effects on the attachment pads and friction forces in syrphid flies (Diptera, Syrphidae). *Journal of Experimental Biology, 204*, 1421-1431.

Gorb, S. (2007). Visualization of native surface by two-step molding. *Microscopy Today*, 44-46.

Gorb, S. N., Beutel, R. G., Gorb, E. V., et al. (2002). Structural design and biomechanics of friction-based releasable attachment devices in insects. *Integrative and Comparative Biology, 42*, 1127-1139.

Gorb, S. N., & Gorb, E. V. (2004). Cntogenesis of the attachment ability in the bug *Coreus marginatus* (Heteroptera, Insecta). *Journal of Experimental Biology, 207,* 2917-2924.

Gosline, J. M., Denny, M. W., & Demont, M. E. (1984). Spider silk as rubber. *Nature, 309,* 551-552.

Gosline, J. M., Demont, E. M., & Denny, M. W. (1986). The structure and properties of spider silk. *Endeavor, 10,* 37-44.

Gosline, J. M., Guerette, P. A., Ortlepp, C. S., et al. (1999). The mechanical design of spider silks: from fibroin sequence to mechanical function. *Journal of Experimental Biology, 202,* 3295-3303.

Gravish, N., Wilkinson, M., & Autumn, K. (2008). Frictional and elastic energy in gecko adhesive.

detachment. *Journal of the Royal Society Interface, 6,* 339-348.

Guduru, P. R. (2007). Detachment of a rigid solid from an elastic wavy surface: theory. *Journal of the Mechanics and Physics of Solids, 55,* 445-472.

Guinea, G. V., Elices, M., Perez-Rigueiro, J., et al. (2003). Self-tightening of spider silk fibers induced by moisture. *Polymer, 44,* 5785-5788.

Haase, A. (1900). Untersuchungen über den bau und die entwicklung der haftlappen bei den geckotiden. *Archiv. Naturgesch, 66,* 321-345.

Haeshin, L., Bruce, P. L., & Phillip, B. M. (2007). A reversible wet/dry adhesive inspired by mussels and geckos. *Nature, 448,* 338-341.

Hajer, J., Maly, J., Hruba, L., et al. (2009). Egg sac silk of *Theridiosoma gemmosum* (Araneae: Theridiosomatidae). *Journal of Morphology, 270,* 1269-1283.

Hansen, W. R., & Autumn, K. (2005). Evidence for self-cleaning in gecko setae. *Proceedings of the National Academy of Sciences, 102,* 385-389.

Hayashi, C. Y., Blackledge, T. A., & Lewis, R. V. (2004). Molecular and mechanical characterization of aciniform silk uniformity of iterated sequence modules in a novel member of the spider silk fibroin gene family. *Molecular Biology and Evolution, 21,* 1950-1959.

He, B., Patankar, N. A., & Lee, J. (2003). Multiple equilibrium droplet shapes and design criterion for rough hydrophobic surfaces. *Langmuir, 19*, 4999-5003.

Herminghaus, S. (2000). Roughness induced non-wetting. *Europhysics Letters, 52*, 165-170.

Hiller, U. (1968). Untersuchungen zum Feinbau und zur Funktion der Haftborsten von Reptilien. *Z. Morphol. Tiere, 62*, 307-362.

Hiller, U. (1969). Correlation between corona-discharge of polyethylene films and the adhering power of *Tarentola in. mauritanica* (Rept.). *Forma et function, 1*, 350-352.

Hora, S. L. (1923). The adhesive apparatus on the toes of certain geckos and tree frogs. *Journal and Proceedings of the Asiatic Society of Bengal, 9*, 137-145.

Hozumi, A., & Takai, O. (1998). Preparation of silicon oxide films having a water-repellent layer by multiple-step microwave plasma-enhanced chemical vapor deposition. *Thin Solid Films, 334*, 54-59.

Huber, G., Mantz, H., Spolenak, R., et al. (2005a). Evidence for capillarity contributions to gecko adhesion from single spatula nanomechanical measurements. *Proceedings of the National Academy of Sciences, 102*, 16293-16296.

Huber, G., Gorb, S. N., Spolenak, R., et al. (2005b). Resolving the nanoscale adhesion of individual gecko spatulae by atomic force microscopy. *Biology Letters, 1*, 2-4.

Huber, G., Gorb, S. N., Hosoda, N., et al. (2007). Influence of surface roughness on gecko adhesion. *Acta Biomaterialia, 3*, 607-610.

Irschick, D. J., Austin, C. C., Petren, K., et al. (1996). A comparative analysis of clinging ability among pad-bearing lizards. *Biological Journal of the Linnean Society, 59*, 21-35.

Irschick, D. J., Vanhooydonck, B., Herrel, A., et al. (2003). Effects of loading and size on maximum power output and kinematics in geckos. *Journal of Experimental Biology, 206*, 3923-3934.

Israelachvili, J. N. (1992). Intremolecular and surface forces. In: 2nd edn. Academic press, London.

Jung, Y.C., & Bhushan, B. (2007a). Contact angle, adhesion and friction properties of micro- and nanopatterned polymers for superhydrophobicity. *Nanotechnology, 17*, 4970-80.

Jung, Y. C., & Bhushan, B. (2007b). Wetting transition of water droplets on superhydrophobic patterned surfaces. *Scripta Materialia, 57*, 1057-60.

Jung, Y. C., & Bhushan, B. (2008). Wetting behavior during evaporation and condensation of water microdroplets on superhydrophobic patterned surfaces. *Journal of Microscopy, 229*, 127-40.

Jusufi, A., Goldman, D. I., Revzen, S., et al. (2008). Active tails enhance arboreal acrobatics in geckos. *Proceedings of the National Academy of Sciences, 105*, 4215-4219.

Kesel, A. B., Martin, A., & Seidl, T. (2003). Adhesion measurements on the attachment devices of the jumping spider *Evarcha arcuata. Journal of Experimental Biology, 206*, 2733-2738.

Kesel, A. B., Martin, A., & Seidl, T. (2004). Getting a grip on spider attachment. *Smart Materials & Structures, 13*, 512-518.

Keten, S., Xu, Z., Ihle, B., et al. (2010a). Nanoconfinement controls stiffness, strength and mechanical toughness of b-sheet crystals in silk. *Nature Materials, 9*, 359-367.

Keten, S., & Buehler, M. J. (2010b). Nanostructure and molecular mechanics of dragline spider silk protein assemblies. *Journal of the Royal Society Interface, 7*, 1709-1721.

Kim, T. W., & Bhushan, B. (2007). Effect of stiffness of multi-level hierarchical attachment system on adhesion enhancement. *Ultramicroscopy, 107*, 902-912.

Koch, K., Dommisse, A., Barthlott, W., et al. (2007). The use of plant waxes as templates for micro- and nanopatterning of surfaces. *Acta Biomaterialia, 3*, 905-909.

Koch, K., Schulte, A. J., Fischer, A., et al. (2008). A fast, precise and low-cost replication technique for nano- and high-aspect-ratio structures of biological ad artificial surfaces. *Bioinspiration and Biomimetics, 3*, 046002 (10 pp).

Koch, K., Bhushan, B., & Barthlott, W. (2009). Multifunctional surface structures of plants: An inspiration for biomimetics. *Progress in Materials Science, 54*, 137-178.

Köhler, T., & Vollrath, F. (1995). Thread biomechanics in the two orb weaving spiders *Araneus diadematus* (Araneae, Araneidae) and *Uloborus walckenaerius* (Araneae, Uloboridae). *Journal of Experimental Zoology, 271*, 1-17.

Kovoor, J. (1987). *Comparative structure and histochemistry of silk-producing organs in arachnids. Ecophysiology of Spiders.* Berlin: Springer-Verlag. 160-186.

Krasnov, I., Diddens, I., Hauptmann, N., et al. (2008). Mechanical properties of silk: interplay of *deformation* on macroscopic and molecular length scales. *Physical Review Letters, 100*, 1-4.

Lau, K. K. S., Bico, J., Teo, K. B. K., et al. (2003). Superhydrophobic carbon nanotube forests. *Nano Letters, 3*, 1701-5.

Lee, J., Fearing, R. S., & Komvopoulos, K. (2008a). Directional adhesion of gecko-inspired angled microfiber arrays. *Applied Physics Letters, 93*, 191910.

Lee, S. M., Lee, H. S., Kim, D. S., et al. (2006a). Fabrication of hydrophobic films replicated from plant leaves in nature. *Surface and Coatings Technology, 201*, 553-9.

Lee, S., & Kwon, T. H. (2006b). Mass-producible replication of highly hydrophobic surfaces from plant leaves. *Nanotechnology, 17*, 3189-3196.

Lee, S. M., & Kwon T. H. (2007). Effects of intrinsic hydrophobicity on wettability of polymer replicas of a superhydrophobic lotus leaf. *Journal of Micromechanics and Microengineering, 17*, 687-92.

Lee, S. M., Jung, I. D., & Ko, J. S. (2008b). The effect of the surface wettability of nanoprotrusions formed on network-type microstructures. *Journal of Micromechanics and Microengineering, 18*, 125007 (7pp).

Lees, A. D., & Hardie, J. (1988). The organs of adhesion in the aphid Megoura viciae. *Journal of Experimental Biology, 136*, 209-228.

Lepore E., Brianza S., Antoniolli F., et al. (2008). Preliminary in vivo experiments on adhesion of geckos. *Journal of Nanomaterials*, Article ID 194524, 5 pages.

Lepore E., Chiodoni A., Pugno N. (2010). New topological and statistical observations on the moult and skin of tokay geckos. *Reviews on Advanced Materials Science, 24*, 69-80.

Lepore, E., Faraldi, P., Boarino, L., et al. (2012a). Plasma and thermoforming treatments to tune the bio-inspired wettability of polystyrene. *Composites Part B: Engineering, 43*, 681-690.

Lepore, E., Marchioro, A., Isaia, M., et al. (2012b). Evidence of the most stretchable egg sac silk stalk, of the European spider of the year Meta Menardi. PLoS ONE, 7, 1-12.

Lepore, E., Pugno, F., & Pugno, N. (2012c). Optimal angles for maximal adhesion in living tokay geckos. Journal of adhesion, 88, 820-830.

Lepore, E., Brambilla, P., Pero, A., et al. (2013). Observations of shear adhesive force and friction of Blatta orientalis on different surfaces. MECCANICA, Special Issue Micro- and Nano-mechanics, 48, 8, 1863-1873. DOI: 10.1007/s11012-013-9796-6. ISSN (print): 0025-6455. ISSN (online): 1572-9648.

Liang, Y. A., Autumn, K., Hsieh, S. T., et al. (2000). *Technical Digest of the 2000 Solid-State Sensor and Actuator Workshop* 200033, 38.

Liu, B., He, Y., Fan, Y., et al. (2006). Fabricating superhydrophobic lotus-leaf-like surfaces through soft-lithographic imprinting. *Macromolecular Rapid Communications, 27*, 1859-64.

Lu-quan, R., Shu-jie, W., Xi-mei, T., et al. (2007). Non-smooth morphologies of typical plant leaf surfaces and their anti-adhesion effects. *Journal of Bionics Engineering, 4*, 33-40.

Maderson, P. F. A. (1964). Keratinized epidermal derivatives as an aid to climbing in gekkonid lizards. *Nature, 203*, 780-781.

Madsen, B., Shao, Z. Z., & Vollrath, F. (1999). Variability in the mechanical properties of spider silks on three levels: interspecific, intraspecific and intraindividual. *International Journal of Biological Macromolecules, 24*, 301-306.

Mahendra, B. C. (1941). Keratinized Epidermal Derivatives as an Aid to Climbing in Gekkonid Lizards. *Proceedings of the Indian Academy of Sciences, 13*, 288-306.

Miwa, M., Nakajima, A., Fujishima, A., et al. (2000). Effects of the surface roughness on sliding angles of water droplets on superhydrophobic surfaces. *Langmuir, 16*, 5754-60.

Moore, A. M. F., & Tran, K. (1999). Material properties of cobweb silk from the black widow spider *Latrodectus Hesperus*. *International Journal of Biological Macromolecules, 24*, 277-282.

Murphy, M. P., Aksak, B., & Sitti, M. (2007). Adhesion and anisotropic friction enhancements of angled heterogeneous micro-fiber arrays with spherical and spatula tips. *Journal of Adhesion Science and Technology, 21,* 1281.

Neinhuis, C., & Barthlott, W. (1997). Characterization and distribution of water-repellent, self-cleaning plant surfaces. *Annals of Botany, 79,* 667-677.

Niederegger, S., & Gorb, S. N. (2006). Friction and adhesion in the tarsal and metatarsal scopulae of spiders. *Journal of Comparative Physiology A, 192,* 1223-1232.

Noeske, M., Degenhardt, J., Strudhoff, S., et al. (2004). Plasma jet treatment of five polymers at atmospheric pressure: surface modifications and the relevance for adhesion. *International Journal of Adhesion and Adhesives, 2,* 171-177.

Nosonovsky, M., & Bhushan, B. (2005). Roughness optimization for biomimetic superhydrophobic surfaces. *Microsystem Technologies, 11,* 535-549.

Nosonovsky, M., & Bhushan, B. (2006). Wetting of rough three-dimensional superhydrophobic surfaces. *Microsystem Technologies, 12,* 273-81.

Nosonovsky, M., & Bhushan, B. (2007a). Hierarchical roughness optimization for biomimetic superhydrophobic surfaces. *Ultramicroscopy, 107,* 969-79.

Nosonovsky, M., & Bhushan, B. (2007b). Biomimetic superhydrophobic surfaces: Multiscale approach. *Nano Letters, 7,* 2633-7.

Nosonovsky, M., & Bhushan, B. (2007c). Multiscale friction mechanisms and hierarchical surfaces in nano-and bio-tribology. *Materials Science and Engineering: R: Reports, 58,* 162-193.

Nosonovsky, M., & Bhushan, B. (2007d). Hierarchical roughness makes superhydrophobic states stable. *Microelectronic Engineering, 84,* 382-6.

Nosonovsky, M., & Bhushan, B. (2008). Patterned non-adhesive surfaces: Superhydrophobicity and wetting regime transitions. *Langmuir, 24,* 1525-1533.

Nova, A., Keten, S., Pugno, N. M., et al. (2010). Molecular and nanostructural mechanisms of deformation, strength and toughness of spider silk fibrils. *Nano Letters, 10,* 2626-2634.

Oner, D., & McCarthy, T. J. (2000). Ultrahydrophobic surfaces. Effects of topography length scales on wettability. *Langmuir, 16*, 7777-7782.

Opell, B. D., & Bond, J. E. (2000). Capture thread extensibility of orb-weaving spiders: testing punctuated and associative explanations of character evolution. *Biological Journal of the Linnean Society, 70*, 107-120.

Ortlepp, C., & Gosline, J. M. (2008). The scaling of safety factor in spider draglines. *Journal of Experimental Biology, 211*, 2832-2840.

Otten, A., & Herminghaus, S. (2004). How plants keep dry: A physicists point of view. *Langmuir, 20*, 2405-8.

Peattie, A. M. (2009). Functiona. demands of dynamic biological adhesion: an integrative approach. *Journal of Comparative Physiology B, 179*, 231-239.

Peressadko, A. G., & Gorb, S. N. (2004). *Surface profile and friction force generated by insects.* In: Boblan I, Bannasch R, editors. Bionik, 15, 237.

Perez-Rigueiro, J., Viney, C., Llorca, J., et al. (1998). Silkworm silk as an engineering material. *Journal of Applied Polymer Science, 70*, 2439-2447.

Perez-Rigueiro, J., Viney, C., Llorca, J., et al. (2000). Mechanical properties of single-brin silkworm silk. *Journal of Applied Polymer Science, 75*, 1270-1277.

Perez-Rigueiro, J., Elices, M., Llorca, J., et al. (2001). Tensile properties of *Argiope trifasciata* drag line silk obtained from the spider's web. *Journal of Applied Polymer Science, 82*, 2245-2251.

Persson, B. N. J., & Tosatti, E. (2001). The effect of surface roughness on the adhesion of elastic solids. *The Journal of Physical Chemistry, 115*, 5597-5610.

Persson, B. N. J. (2002). Adhesion between and elastic body and a randomly hard surface. *European Physical Journal E, 8*, 385-401.

Persson, B. N. J., & Gorb S. (2003). The effect of surface roughness on the adhesion of elastic plates with application to biological systems. *The Journal of Physical Chemistry, 119*, 11437-11444.

Persson, B. N. J. (2007). Biological adhesion for locomotion on rough surfaces: basic principles and a theorist's view. *Mrs Bulletin, 32*, 486-490.

Pesika, N. S., Tian, Y., Zhao, B., et al. (2007). Peel-zone model of tape peeling based on the gecko adhesive system. *Journal of Adhesion, 83*, 383-401.

Platnick, N. I. (2011). The world spider catalogue. version 12.0. American Museum of Natural History.

Poza, P., Perez-Rigueiro, J., Elices, M., et al. (2006). Fractographic analysis of silkworm and spider silk. *Engineering Fracture Mechanics, 69*, 1035-1048.

Pugno, N. (2006). Mimicking nacres width super-nanotubes for producing optimized super-composites. *Nanotechnology, 17*, 5480-5484.

Pugno, N. M. (2007a). Towards a Spiderman suit: large invisible cables and self-cleaning releasable superadhesive materials. *Journal of Physics: Condensed Matter, 19*, 395001 (17pp).

Pugno, N. M. (2007b). *The Nanomechanics in Italy*, Research signpost (IND).

Pugno N., Lepore E. (2008a). Living tokay geckos display adhesion times following the Weibull Statistics. *Journal of Adhesion, 84*, 949-962.

Pugno N., Lepore E. (2008b). Observation of optimal gecko's adhesion on nanorough surfaces, *Biosystems, 94*, 218-222.

Pugno, N. (2008c). Spiderman gloves. *Nano Today, 3*, 35-41.

Pugno, N., Bosia, F., & Carpinteri, A. (2008d). Multiscale stochastic simulations for tensile testing of nanotube-based macroscopic cables. *Small, 4*, 1044-1052.

Pugno, N., & Carpinteri, A. (2008e). Design of micro-nanoscale bio-inspired hierarchical materials. *Philosophical Magazine Letters, 88*, 397-405.

Pugno, N. M. (2010). The design of self-collapsed super-strong nanotube bundles. *Journal of the Mechanics and Physics of Solids, 58*, 1397-1410.

Pugno N., Lepore, E., Toscano, S., & Pugno, F. (2011). Normal adhesive force-displacement curves of living geckos. *Journal of Adhesion, 87*, 1059-1072.

Pugno, N. (2011). *International Journal of Fracture*, ICF XII Special Issue on Nanoscale Fracture, Guest Editor Nicola M. Pugno (2011). In Print. (published as arXiv: 0903.0935 cond-mat, 5 march 2009). Available online from the 14 November 2011, http://www.springerlink.com/content/um36255632007747/fulltext.pdf.

Quéré, D. (2002). Fakir droplets. *Nature Materials, 1*, 14-5.

Raibeck, L., Reap, J., & Bras, B. (2008). Investigating environmental benefits of
biologically inspired self-cleaning surfaces.15th CIRP International Conference
on Life Cycle Engineering.

Römer, L., & Scheibel, T. (2008). The elaborate structure of spider silk. *Prion, 4*,
154-161.

Rousseau, M. E., Lefevre, T., & Pezolet, M. (2009). Conformation and orientation
of proteins in various types of silk fibers produced by Nephila clavipes Spiders.
Biomacromolecules, 10, 2945-2953.

Ruibal, R., & Ernst, V. (1965). The structure of the digital setae of lizards. *Journal
of Morphology, 117*, 271-293.

Russell, A. P. (1975). A contribution to the functional morphology of the foot of
the tokay, *Gekko gecko. Journal of Zoology, 176*, 437-476.

Russell, A. P. (1986). The morphological basis of weight-bearing in the scansors
of the tokay gecko (*Gekko gecko*). *Canadian Journal of Zoology, 64*, 948-955.

Russell, A. P. (2002). Integrative functional morphology of the gekkotan
adhesive system (Reptilia: Gekkota). *Integrative and Comparative Biology, 42*,
1154-1163.

Russell A. P., & Higham, T. E. (2009). A new angle on clinging in geckos: Incline,
not surface structure, triggers the deployment of adhesive system. *Proceedings
of Biological Sciences, 276*, 3705-3709.

Santos, D., Spenko, M., Parness A., et al. (2007). Directional Adhesion for
Climbing: Theoretical and Practical Considerations. *Journal of Adhesion Science
and Technology, 21*, 1317.

Schleich, H. H., & Kästle, W. (1986). Ultrastrukturen an Gecko-Zehen (Reptilia:
Sauria: Gekkonidae). *Amphibian & reptile, 7*, 141-166.

Schmidt, H. R. (1904). *Jena. Z. Naturwiss, 39*, 551-580.

Schubert, B., Lee, J., Majidi C., et al. (2008). Sliding-induced adhesion of stiff
polymer microfibre arrays. *Journal of the Royal Society Interface, 5*, 845-853.

Schulte, A. J., Koch, K., Spaeth, M., et al. (2009). Biomimetic replicas: Transfer of complex architectures with different optical properties from plant surfaces onto technical materials. *Acta Biomaterialia, 5*, 1848-1854.

Sen, D., Novoselov, K., Reis, P., et al. (2010). Tearing of graphene sheets from adhesive substrates produces tapered nanoribbons. *Small, 6*, 1108-1116.

Sensenig, A., Agnarsson, I., & Blackledge, T. A. (2010). Behavioural and biomaterial coevolution in spider orb webs. *Journal of Evolutionary Biology, 23*, 1839-1856.

Shah, G. J., & Sitti, M. (2004). Modeling and design of biomimetic adhesives inspired by gecko foot-hairs. *Proceedings of the 2004 IEEE, International Conference on Robotics and Biomimetics*, August 22 - 26, Shenyang, China.

Shao, Z., & Vollrath, F. (2008). The effect of solvents on the contraction and mechanical properties of spider silk. *Polymer, 40*, 1799-1806.

Shibuichi, S., Onda, T., Satoh, N., et al. (1996). Super-waterrepellent surfaces resulting from fractal structure. *Journal of Physical* Chemistry, 100, 19512-7.

Simmermacher, G. (1884). Untersuchungen über die Haftapparate an Tarsalgliedern von Insekten. *Zeitschr. Wiss. Zool, 40*, 481-556.

Singh, R. A., Yoon, E., Kim, H. J., et al. (2007). Replication of surfaces of natural leaves for enhanced micro-scale tribological property. *Materials Science and Engineering: C, 27*, 875-879.

Sitti, M., & Fearing, R. S. (2003). Synthetic gecko foot-hair micro/nano structures as dry adhesives. *Journal of Adhesion Science and Technology, 17*, 1055-1073.

Solga, A., Cerman, Z., Striffler, B. F., et al. (2007). The dream of staying clean: Lotus and biomimetic surfaces. *Bioinspiration and Biomimetics, 2*, 126-134.

Spolenak, R., Gorb, S. & Arzt, E. (2005). Adhesion design maps for bio-inspired attachment systems. *Acta Biomaterialia, 1*, 5-13.

Stauffer, S. L., Coguill, S. L., & Lewis, R. V. (1994). Comparison of physical properties of three silks from *Nephila clavipes* and *Araneus gemmoides*. *Journal of Arachnology, 22*, 5-11.

Stork, N. E. (1980). Experimental analysis of adhesion of *Chrysolina polita* (Chrysomelidae: Coleoptera) on a variety of surfaces. *Journal of Experimental Biology, 88,* 91-107.

Sun, W., Neuzil, P., Kustandi, T. S., Oh, S., & Samper, V. D. (2005a). The nature of the gecko lizard adhesive force. *Biophysical Journal, 89,* L14-L17.

Sun, M., Luo, C., Xu, L., et al. (2005b). Artificial Lotus leaf by nanocasting. *Langmuir, 21,* 8978-8981.

Swanson, B. O., Blackledge, T. A., Beltran, J., et al. (2006). Variation in the material properties of spider dragline silk across species. *Applied Physics A, 82,* 213-218.

Tian, Y., Pesika, N., Zeng, H., et al. (2006). *Proceedings of the National Academy of Sciences, 103,* 19320-19325.

Tinkle, D. W. (1992). *Gecko.* In: Cummings M (ed) Encyclopedia Americana, vol 12. Grolier, London, p 359.

Valeri, S., Paradisi, P., Luches, P., et al. (1999). Growth and morphology of Te films on Mo. *Thin Solid Films, 352,* 114-118.

Van Casteren, A., & Codd, J. A. (2008). Foot morphology and substrate adhesion in the Madagascan hissing cockroach, *Gromphadorhina portentosa. Journal of Insect Science, 10,* 1-11.

Van Nimmen, E., Gellynck, K., Van Langenhove, L., et al. (2006). The tensile properties of cocoon silk of the spider *Araneus diadematus. Textile Research Journal, 76,* 619-628.

Van Nimmen, E., Gellynck, K., & Van Langenhove, L. (2005a). The Tensile Behaviour of Spider Silk. *Autex Research Journal, 5,* 120-126.

Van Nimmen, E., Gellynck, K., Gheysens, T., et al. (2005b). Modelling of the Stress-Strain behaviour of egg sac silk of the spider *Araneus diadematus. Journal of Arachnology, 33,* 629-639.

Varenberg, M., Pugno, N., & Gorb, S. (2010). Spatulate structures in biological fibrillar adhesion. *Soft Matter, 6,* 3269-3272.

Vasanthavada, K., Hu, X., Falick, A. M., et al. (2007). Aciniform spidroin, a constituent of egg case sacs and wrapping silk fibers from the black widow spider *Latrodectus Hesperus*. *The Journal of Biological Chemistry, 282*, 35088-35097.

Vehoff, T., Glišović, A., Schollmeyer, H., et al. (2007). Mechanical properties of spider dragline silk: Humidity, hysteresis and relaxation. *Biophysical Journal, 93*, 4425-4432.

Voigt, D. (2005). Dissertation. Technische Universität Dresden, Dresden, Germany, 185 pp.

Voigt, D., Schuppert, J. M., Dattinger, S., & Gorb, S. N. (2008). Sexual dimorphism in the attachment ability of the Colorado potato beetle *Leptinotarsa decemlineata* (Coleoptera: Chrysomelidae) to rough substrates. *Journal of Insect Physiology, 54*, 765-776.

Vollrath, F., Holtet, T., Thorgensen, H. C., et al. (1996). Structural organization of spider silk. *Proceedings of the Royal Society B: Biological Sciences, 263*, 147-151.

Vollrath, F., Madsen, B., & Shao, Z. (2001). The effect of spinning conditions on the mechanics of a spider's dragline silk. *Proceedings of the Royal Society of London Series B, 268*, 2339-2346.

Wagler, J. J. G. (1830). Natürliches System der Amphibien, mit Vorangehander Classification der Säugtiere und Vögel. *J. G. Cotta'schen Buchhandlung*, München. 354 pp.

Wagner, P., Furstner, R., Barthlott, W., et al. (2003). Quantitative assessment to the structural basis of water repellency in natural and technical surfaces. *Journal of Experimental Botany, 385*, 1295-1303.

Walker, G., Yue, A. B., & Ratcliffe, J. (1985). The adhesive organ of the blowfly, *Calliphora vomitoria*: a functional approach (Diptera: Calliphoridae). *Journal of Zoology London A, 205*, 297-307.

Walker, G. (1993). Adhesion to smooth surfaces by insects-a review. *International Journal of Adhesion and Adhesives, 13*, 3-7.

Weitlaner, F. (1902). Eine Untersuchung über den Haftfuß des Gecko. *Verhdl. Zool. Bot. Ges. Wien, 52*, 328-332.

Wenzel, R. N. (1936). Resistance of solid surfaces to wetting by water. *Industrial and Engineering Chemistry, 28*, 988-94.

Wigglesworth, V. B. (1987). How does a fly cling to the under surface of a glass sheet? *Journal of Experimental Biology, 129,* 373-376.

Williams, E. E., & Peterson, J. A. (1982). Convergent and alternative designs in the digital adhesive pads of scincid lizards. *Science, 215,* 1509-1511.

Wu, X., Zheng, L., & Wu, D. (2005). Fabrication of superhydrophobic surfaces from microstructured ZnO based surfaces via a wet-chemical route. *Langmuir, 21,* 2665-7.

Xiao, T., Ren, Y., Liao, K., et al. (2008). Determination of tensile strength distribution of nanotubes from testing of nanotube bundles. *Composites Science and Technology, 68,* 2937-2942.

Yao, H., & Gao, H. (2006). Mechanics of robust and releasable adhesion in biology: bottom-up designed hierarchical structures of gecko. *Journal of the Mechanics and Physics of Solids, 54,* 1120-1146.

Yao, H., Rocca, G. D., Guduru, P. R., et al. (2008). Adhesion and sliding response of a biologically inspired fibrillar surface: Experimental observations. *Journal of the Royal Society Interface, 5,* 723.

Yeo, J., Choi, M. J., & Kim, D. S. (2010). Robust hydrophobic surfaces with various micropillar arrays. *Journal of Micromechanics and Microengineering, 20,* 025028 (8 pp).

Young, T. (1805). An essay on the cohesion of fluids. *Philosophical Transactions of the Royal Society, 95,* 65-87.

Yuan, Z., Chen, H., Tang, J., et al. (2007). Facile method to fabricate stable superhydrophobic polystyrene surface by adding ethanol. *Surface and Coatings Technology, 201,* 7138-7142.

Yurdumakan, B., Raravikar, N. R., Ajayan, P. M., et al. (2005). Synthetic gecko foot-hairs from multiwalled carbon nanotubes. *Chemical Communications, 30,* 3799-3801.

Zaaf, A., Van Damme, R., Herrel, A., et al. (2001). Spatio-temporal gait characteristics of level and vertical locomotion in aground-dwelling and a climbing gecko. *Journal of Experimental Biology, 204,* 1233-1246.

Zhai, L., Cebeci, F. C., Cohen, R. E., et al. (2004). Stable superhydrophobic coatings from polyelectrolyte multilayers. *Nano Letters, 4,* 1349.

Zhao, A. C., Zhao, T. F., Nakagaki, K., et al. (2006). Novel molecular and mechanical properties of egg case silk from wasp spider, *Argiope bruennichi*. *Biochemistry, 45*, 3348-3356.

Zhiqing, Y., Hong, C., Jianxin, T., et al. (2007). A novel preparation of polystyrene film with a superhydrophobic surface using a template method. *Journal of Physics D: Applied Physics, 40*, 3485-3489.

Final Acknowledgments

The authors would like to thank Matteo Biasotto of the Department of Special Surgery of the University of Trieste for experimental instruments of surface measurements (profilometer and AFM), and Francesca Antoniolli of the Department of Biomedicine, Unit of Dental Sciences and Biomaterials of the University of Trieste, for the helpful scientific discussion. The authors are grateful to Michele Buono and Simona Toscano, DVM and SIVAE member, for the technical and veterinary aid and also for the fundamental support for experimental studies, the biologist Marco Isaia, Department of Life Sciences and Systems Biology of the University of Torino, and Andrea Marchioro of Politecnico di Torino for their help in achieving the experimental tensile tests on the silk bundles of the cocoons of the cave spider Meta menardi, the entomologist Franco Casini for his advice and for providing living cockroaches, Maddalena Binda of the LabSamp of Politecnico di Milano for her helpfulness in surface AFM analysis, Alessandro Pero and Pietro Brambilla of Politecnico di Torino for their help in order to build the centrifuge machine and to perform experiments on the living cockroaches Blatta orientalis, respectively.

We gratefully acknowledge the "213T Scarl - Incubatore dell'Università di Torino" for SEM imaging instruments and Maria Giulia Faga, CNR-ISTEC member, Chemical Department IFM and NIS Centre of Excellence, University of Torino for the fundamental help performing the SEM micrographs of gecko. We thank Luca Boarino and the "Nanofacility Piemonte" for the FESEM imaging instruments and Emanuele Enrico, INRIM Institute, and Angelica Chiodoni, IIT Institute, for the fundamental help performing the FESEM micrographs.

The living plant was provided by the Giardino Botanico Rea (Turin), associated with the Natural Science Museum of Turin, and the authors wish to sincerely thank Rosa Camoletto for this.

A final special thanks goes to Paolo Faraldi and Dino Bongini of Indesit Company for the challenging scientific and industrial collaboration for the fabrication of superhydrophobic and self-cleaning surfaces.

General Conclusions

Over the past 150 years, scientific researchers have examined the adhesive abilities of insects, spiders, reptiles, and geckos by examining their adhesive systems, climbing abilities, and adhesion mechanisms. Moreover, the anti-adhesive properties of some plants have attracted interest in understanding superhydrophobicity and self-cleaning properties. These properties are related to the anti-adhesion of water and dirt. Finally, the strong behaviour of spider silk has suggested interesting applications because of its structural and mechanical properties.

In nature, the common key of adhesive (Chapters 1-5), antiadhesive (Chapters 6-7) and strong (in Chapter 8) properties of materials exists at the nanoscale: from the super-adhesive terminal unit contacts of geckos (spatulae, ~200 nm wide and 15-20 nm thick) to the finest super-anti-adhesive structure of lotus leaves (nanotubules, diameter of ~200 nm) to the fibroin protein of spider silks.

In Chapter 1, we demonstrate that living Tokay geckos (*Gekko gecko*) display adhesion times following Weibull statistics. The Weibull shape (m) and scale (t_0) parameters quantitatively describe the statistics of adhesion times of different geckos (male and female) on glass and Poly(methyl meth-acrylate) (PMMA) ($m_{PMMA} \approx 1$ and $t_{0PMMA} \approx 800$ s *versus* $m_{Glass} \approx 2$ and $t_{0Glass} \approx 23$ s).

Chapter 2 confirms that the Weibull modulus has a value in the restricted range of 1-1.2 when both virgin and machined PMMA surfaces are considered.

Chapter 3 highlights the normal adhesive abilities of living Tokay geckos adhering to PMMA and glass surfaces. The normal safety factor λ, the ratio between the maximum normal adhesive force and the body weight, was thus determined as $\lambda_{PMMA} = 10.23$ on PMMA surface and $\lambda_{Glass} = 9.13$ on glass surface. In addition, the self-renewal of the gecko's adhesive system after moulting was documented.

Chapter 4 investigates the adhesion angles of living Tokay geckos at two different hierarchical levels of the feet and toes. The adhesion angles between opposing front and rear feet (β_F) and between the first and fifth toe of each foot (β_T) on different inverted surfaces (steel, aluminium, copper, Poly(methyl meth-acrylate), and glass) have been experimentally measured. The resulting angle α was computed as $\alpha= (180°-\beta)/2$ and found to be to 28° (α_{F_FR-RL}) and 30° (α_{F_FL-RR}) for the opposing front and rear feet and 26° (α_{T_FR}), 29° (α_{T_FL}), 28° (α_{T_RR}), and 26° (α_{T_RL}) between the first and fifth toe of each foot. Such results are consistent with the recently described multiple peeling theory: as the number of hierarchical level n increases, the dimensionless adhesion strength parameter λ decreases and determines a decrease of the adhesion angle α.

Chapter 5 ends the first section of this study on adhesive materials. In this chapter, the shear adhesive force of four non-climbing living cockroaches (*Blatta Orientalis* Linnaeus) was investigated using a centrifuge machine on six surfaces (steel, aluminium, copper, two sand papers (Sp 50, Sp150), and a common paper sheet). The shear safety factor was determined as the ratio between the maximum shear adhesive force and the body weight: the cockroach's maximum shear adhesive factor is 12.1 or Sp150, whereas the minimum shear adhesive factor is 1.9 on steel surface.

Chapter 6 displays the effects of two superficial industrial processes (plasma and thermoforming) on the surface wettability of polystyrene (PS). This analysis was developed in collaboration with the Indesit Company and suggests that plasma and thermoforming are ideal treatments to tune the wettability and enhance hydrophilic or hydrophobic behaviour of PS surfaces, respectively.

Chapter 7 shows how an artificial biomimetic hydrophobic polystyrene (PS) surface can be constructed by copying a natural lotus leaf with a simple template method at room temperature and atmospheric pressure. Two parameters were used to compare the artificial PS surface *vs* the natural lotus leaf: the contact angle, of 149° *vs.* 153°, and the drop sliding speed, of 417 mm/s *vs.* 319.4 mm/s, respectively.

Chapter 8 is concerns the natural stretchable egg sac silk stalks of the cave spider *Meta menardi*. Bundles of about 150 threads (each of ~6 μm in diameter) connecting the egg sac (cocoon) to the cave ceilings were tested to determine stress-strain curves, and the stress results were analysed with Weibull statistics. The maximum stress, strain, and toughness modulus reach the values of 0.64 GPa, 750 %, and 130.7 MJ/m³, respectively. The average value of the Weibull modulus (m) is in the range 1.5-1.8, and that of the Weibull scale parameter (σ_0) is in the range 0.33-0.41 GPa with a high correlation coefficient (R^2 = 0.97).

As the mechanisms in nature governing adhesive, anti-adhesive, and strong properties become better understood, these mechanisms will be imported into human technology. A super-adhesive and reusable material has great potential for space applications (*i.e.* connections between space components or suits and gloves for astronauts, which would allow them to remain attached to the external side of a space shuttle without awkward cables); a super-hydrophobic and self-cleaning material could have great utility for civil engineering applications (*i.e.* the glass windows of skyscrapers or external building coverings) or even air transport security (*i.e.* a superficial pattern to anti-ice airplane wings) or home technology (*i.e.* the internal faces of refrigerators or freezers and the surfaces of bathroom fittings or tiles). Finally, a super-stretchable material has potential industrial applications for air transport security (*i.e.* security systems to decrease the velocity of an airplane or large webs to stop hijacked airplanes).

Although efforts to industrialize products like those mentioned above are considerable, only a few patents have been duly deposited at the European Patent Office in these fields of research. In particular, the number of patents with the words "super adhesive reusable" as title keywords is equal to zero, whereas there are eight patents for inventions with the title keywords "self cleaning super hydrophobic" during the last eight years (i.e. a current rate of one European patent per year) and only one patent with the keyword "super stretchable" has been deposited in 1987.

As we survey, the current cutting-edge technology, we understand the multitude of possibilities that exist to become Innovators who develop new products in these fields of scientific research. Nevertheless, only a deep and detailed knowledge of what happens in nature and the understanding of how nature has optimized each process, mechanism, and animal to its own habitat will allow the development of new engineered products.

Index